Communications in Computer and Information Science   1722

More information about this series at https://link.springer.com/bookseries/7899

Johannes Josef Schneider ·
Mathias Sebastian Weyland · Dandolo Flumini ·
Rudolf Marcel Füchslin (Eds.)

Artificial Life and Evolutionary Computation

15th Italian Workshop, WIVACE 2021
Winterthur, Switzerland, September 15–17, 2021
Revised Selected Papers

 Springer

Editors
Johannes Josef Schneider ⓘ
Zurich University of Applied Sciences
Winterthur, Switzerland

School of Engineering
Cardiff University
Cardiff, UK

Dandolo Flumini ⓘ
Zurich University of Applied Sciences
Winterthur, Switzerland

Mathias Sebastian Weyland ⓘ
Zurich University of Applied Sciences
Winterthur, Switzerland

Rudolf Marcel Füchslin ⓘ
Zurich University of Applied Sciences
Winterthur, Switzerland

European Centre for Living Technology
Venice, Italy

ISSN 1865-0929   ISSN 1865-0937 (electronic)
Communications in Computer and Information Science
ISBN 978-3-031-23928-1   ISBN 978-3-031-23929-8 (eBook)
https://doi.org/10.1007/978-3-031-23929-8

This Springer imprint is published by the registered company Springer Nature Switzerland AG
The registered company address is: Gewerbestrasse 11, 6330 Cham, Switzerland

Preface

This volume of Communication in Computer and Information Science contains the proceedings of WIVACE 2021, the 15th Workshop on Artificial Life and Evolutionary Computation. The event was originally planned to be held at the campus in Winterthur, Switzerland of the Zurich University of Applied Sciences in September 2020, but had to be rescheduled to September 15–17, 2021, and held in a hybrid way because of the COVID-19 pandemic.

The International Workshop on Artificial Life and Evolutionary Computation aims at bringing together researchers working in artificial life and evolutionary computation to present and share their research in a multidisciplinary context. The workshop provides a forum for the discussion of new research directions and applications in these fields. It was first held in 2007 in Sampieri (Ragusa), as the incorporation of two previously separately running workshops (WIVA and GSICE). After the success of the first edition, the workshop was organized in the following years in Venice (2008), Naples (2009), Parma (2012), Milan (2013), Vietri (2014), Bari (2015), Salerno (2016), Venice (2017), Parma (2018), and Rende (2019). In 2021, it was for the first time held outside Italy, demonstrating that WIVACE is not only an Italian but also an international community.

The various contributions gathered a wide range of interdisciplinary topics: (I) Networks, (II) Droplets, Fluids, and Synthetic Biology, (III) Robot Systems, (IV) Computer Vision and Computational Creativity, (V) Semantic Search, (VI) Artificial Medicine and Pharmacy, (VII) Trade and Finance. Furthermore, a panel discussion on (VIII) Ethics in Computational Modeling and a session on Social Systems was held. Overall, 10 invited and 22 contributed talks were given. In total, 23 high-quality papers were accepted for this proceedings volume after a single-blind review round performed by Program Committee members, with each paper receiving at least three reviews. Submissions and participants came from nine different European countries and from Iraq, Japan, and the USA, making WIVACE 2021 a truly international event.

Many people contributed to this successful edition. We express our gratitude to the authors for submitting their works, to the members of the Program Committee for devoting so much effort to reviewing papers, to the invited speakers and the members of the panel discussion about ethics, and to Ramvijay Subramani at Springer for their support through the final stages leading to this proceedings volume. Our gratitude also goes to the Zurich University of Applied Sciences for offering the venue for the event. We would like to thank various members of the staff at the Zurich University of Applied Sciences for helping us with the workshop administration. Hereby, Teresa D'Onghia and Julia Obst deserve special thanks. We would also like to mention Roberto Serra for providing both constructive criticism and inspiration.

November 2022

Johannes Josef Schneider
Mathias Sebastian Weyland
Dandolo Flumini
Rudolf Marcel Füchslin

In Memoriam

We mourn the loss of two members of our WIVACE community, Orazio Miglino (1963–2021) and Domenico Parisi (1935–2021).

Orazio Miglino was the organizer of WIVACE 2009 in Naples.

Domenico Parisi's plenary talk was one of the highlights at WIVACE 2012 in Parma.

Organization

General Chairs

Johannes Josef Schneider   Zurich University of Applied Sciences,
Switzerland

Rudolf Marcel Füchslin   Zurich University of Applied Sciences,
Switzerland

Local Chairs

Dandolo Flumini   Zurich University of Applied Sciences,
Switzerland

Mathias Sebastian Weyland   Zurich University of Applied Sciences,
Switzerland

Program Committee

Alberto Acerbi   Brunel University London, UK
Marco Antoniotti   University of Milano-Bicocca, Italy
Michele Amoretti   University of Parma, Italy
Luca Ascari   University of Parma, Italy
Marco Baioletti   University of Perugia, Italy
Vito Antonio Bevilacqua   Politecnico di Bari, Italy
Eleonora Bilotta   University of Calabria, Italy
Leonardo Bocchi   University of Firenze, Italy
Andrea Bracciali   University of Stirling, UK
Michele Braccini   University of Bologna, Italy
Marcello Budroni   University of Sassari, Italy
Stefano Cagnoni   University of Parma, Italy
Angelo Cangelosi   University of Manchester, UK
Timoteo Carletti   University of Namur, Belgium
Mauro Castelli   Universitade Nova de Lisboa, Portugal
Antonio Chella   University of Palermo, Italy
Franco Cicirelli   University of Calabria, Italy
Antonio Della Cioppa   University of Salerno, Italy
Alessia Faggian   University of Trento, Italy
Harold M. Fellermann   Newcastle University, UK
Francesco Fontanella   Università di Cassino e del Lazio Meridionale,
Italy

Contents

Robot Systems

Computer Vision and Computational Creativity

Networks

Dynamical Criticality in Growing Networks

Giovanni Cappelletti[1], Gianluca D'Addese[1] , Roberto Serra[1,2,3] ,
and Marco Villani[1,2(✉)]

[1] Department of Physics, Informatics and Mathematics, Modena and Reggio Emilia University,
Modena, Italy
254809@studenti.unimore.it, rserra@unimore.i0074,
marco.villani@unimore.it
[2] European Centre for Living Technology, Venice, Italy
[3] Institute of Advanced Studies, University of Amsterdam, Amsterdam, The Netherlands

Abstract. The principle of dynamical criticality is a very important hypotheses in biology, and it therefore deserves a thorough investigation. Testing the principle in real biological cases can be far from trivial: therefore, in this work we make use of the Random Boolean Network framework, which has been extensively used to model genetic regulatory networks, and which has since become one of the most used models in the field of complex systems. We subject several RBN ensembles to evolutionary changes: the key research questions are whether initially critical networks will grow faster than ordered or chaotic ones, and whether evolution can influence the dynamic regime, and in which direction. The results obtained so far indicate that critical systems perform well in the analyzed tasks. In the case of two connections per node, the best performances are those of critical systems, while increasing the value of the connectivity there seems to be a slight shift towards more disordered regimes (albeit still close to the critical one).

Keywords: RBN · Evolution · Critical dynamic regime

1 Introduction

The principle of dynamical criticality [1–6] is one of the few broadly applicable hypotheses in biology. It claims that systems which are in certain types of dynamical regimes (the critical states) display properties which give them an edge with respect to other systems. The dynamically critical states are, loosely speaking, those which are neither too ordered not too disordered. If we consider the asymptotic dynamics of dissipative autonomous deterministic systems, the ordered regime corresponds to constant or cyclic attractors, while disordered states are associated to chaotic attractors. Indeed, dynamically critical states have sometimes been claimed to be "at the edge of chaos" [3].

The qualitative reason why critical states should be advantaged is based on the limited controllability of truly chaotic systems on one side, and on the limited dynamical repertoire of fixed points and limit cycles on the other side. A straightforward consequence of the principle in biology is that evolution should have driven organisms (but possibly also organs and ecosystems) towards critical regimes. However, testing the principle in

J. J. Schneider et al. (Eds.): WIVACE 2021, CCIS 1722, pp. 3–13, 2022.
https://doi.org/10.1007/978-3-031-23929-8_1

real biological cases can be far from trivial. Indications have been obtained e.g. in the case of gene regulatory networks by comparing the properties of real biological systems with those of stylized models, finding that critical or close-to-critical states can provide satisfactory agreements [7–11]. For a broad analysis of the evidence in favor of critical states in biology see [12, 13].

There is nothing in the principle which limits its applicability to biological cases: the advantage of critical states should be observed also in artificial systems (like e.g. robots or simulation models) which can be subject to a more thorough investigation. For the sake of definiteness, from now on we will focus on the Random Boolean Network (RBN) model of gene regulatory networks [1], which has been extensively described in detail elsewhere (see e.g. [2, 3, 14, 15]).

It has been possible to compare the capability of different types of RBNs in different tasks like e.g. reaching an attractor within a predefined number of steps. A number of experiments have proven that one cannot in general claim that critical nets outperform the others [16]. The relative performances may well depend upon the features of the task. This should not be surprising, since the supposed advantage of critical networks should become apparent in changing environments, rather than in fixed tasks. Moreover, sometimes the properties of the evolutionary algorithm may change the bias of the Boolean functions (i.e., the fraction of 1s in their truth table) in a way which in turn may affect the dynamics. The problem then becomes that of identifying under which conditions evolution leads to critical networks.

Previous studies have indeed shown the possibility of achieving Boolean Networks (BNs) with some desired characteristics by means of evolutionary techniques [17–19]. Some of these studies explicitly deal with the relationship between evolution and dynamical regime in the RBN [20], sometimes also forcing the presence of particular dynamical regimes [21, 22].

An interesting alternative to predefined fitness functions is that of directly modelling the process of change of a genetic network, imitating in a stylized way the well-known biological process of gene duplication followed by possible divergence of the new copy from the behavior of its parent [15]. The addition of a new node can be detrimental for the behavior of the system, so it has been proposed to accept the modification iff the new network maintains the old attractors (i.e. the previous "behaviors"), while adding new ones. This choice reflects the hypothesis of a living system already adapted to its environment (by expressing appropriate behaviors through its attractors), and which therefore has an advantage in maintaining them, possibly adding new ones [15].

In this case, the key research questions are whether initially critical networks will grow faster than ordered or chaotic ones, and whether different network types will evolve towards criticality. Preliminary results by Aldana [15] provided clues to a positive answer to the first of these questions, thus supporting the principle of dynamical criticality.

Given the importance of this principle, we performed further tests, broadening the analysis of previous works [15]. In this paper we show the main results obtained so far. They indicate that critical networks perform well in the tasks which have been studied. In the case of two connections per node, the best performances are those of critical networks, while increasing the value of the connectivity there seems to be a slight shift in the optimal value.

2 RBN

A Random Boolean Network (RBN for short) is a dynamical system where time and states are discrete. It was ideated in 1969 [1] for the purpose of investigating genetic regulatory networks, and has proved to be one of the most successful models in the study of complex systems.

The model consists of a graph composed of N nodes, which can take either the value 0 or 1; the relationships between the nodes are represented by directed links. Each node is associated with a dynamic rule, represented by a Boolean table, that provides the correct output to each input combination. In a classical RBN each node has the same number of incoming connections k: the k input nodes are randomly chosen with uniform probability among the remaining $N - 1$ nodes. In absence of detailed biological information, the Boolean function that controls the state of the node can be constructed in different ways. The two main approaches consist in randomly choosing Boolean functions with uniform probability, or in randomly inserting in each output of the Boolean table representing the function the symbol "1" with probability (bias) b and "0" otherwise (the strategy chosen in this paper).

There is no difference in the behavior of the network between choosing a bias b and its opposite $1 - b$, because the symbols 0 and 1 do not have particular semantic meaning. The RNBs used in this paper are synchronous, which means that the change of node states occurs simultaneously for all. The value of a node at time t is determined only by the value of the input nodes at time $t - 1$ and its Boolean function.

The behavior of a RBN can be classified in three different regimes: ordered, critical and chaotic. These regimes differ in their robustness to perturbations. In randomly generated critical networks there is a link between the mean connectivity k and the bias b [14, 23]

$$\frac{1}{k} = 2b(1 - b) \tag{1}$$

so it is possible to determine the regime to which the network belongs. it is useful to note in case of $k = 2$ the chaotic regime does not exist, and that the critical regime is found at $b = 0.5$. In case of $k = 3$ the critical regime is found in $b = 0.21$ (symmetrically in $b = 0.79$); below $b = 0.21$ or above $b = 0.79$ the ordered regime is present, whereas in the range $b \in [0.21, 0.79]$ the chaotic regime is present. In ordered regimes, externally induced temporary disturbances tend to disappear; in chaotic regimes they tend to amplify. An interesting hypothesis assumes that living systems have dynamical regimes close to the separator between order and chaos [2, 3, 24] (or possibly in a volume of phase space close to this surface [25]). Over time some groups have found interesting clues to this effect [7–9].

In order to be able to determine the regime of the system under consideration one must measure the spreading of perturbations through the RBN. The dynamical state is usually identified by considering the behavior of a small perturbation of the state (i.e. by the so-called Derrida parameter, the discrete analogue of the Lyapunov exponent of continuous dynamical systems [23, 26–28]): if the size of the perturbation increases the network is in a chaotic regime ($\lambda > 1$), while if it shrinks it is ordered ($\lambda < 1$). In critical networks, where $\lambda = 1$, the average size of the perturbation stays constant.

3   Evolution

An interesting area of study on RBNs concerns their evolution, which can be achieved through the application of Evolutionary Algorithms. Among them, in this paper Genetic Algorithms (GA in the following) [30] will be considered. Indeed, genetic algorithms have been successfully applied to evolve RBNs which are able to solve some tasks [16, 30, 31] or endowed with particular properties, like robustness [17, 18] or ability to differentiate [19]. An interesting way of applying this approach - which we will follow in this paper - is presented in [15, 32].

In order to apply the GA to RBNs, we need the list of used mutation, the strategy followed for their application, and the definition of the fitness.

Regarding mutations, follows a list of moves focused on a 'local' action on individual nodes, borrowed from what happens in genetic regulatory networks:

- **Boolean function change**: the Boolean function is randomized, by maintaining the same bias b
- **delete an input link**: one of the incoming connections is eliminated; the Boolean function halves its size
- **delete an output link**: one of the outgoing connections is eliminated; in this case the Boolean function of the node that link was pointing to, halves its truth table
- **add an input link**: in this case the Boolean function doubles the size of its truth table
- **add an output link**: in this case the Boolean function of the node that link was pointing to doubles the size of its truth table.

Whenever a link is removed or added, arises the need of modifying the size of the truth table of nodes whose output was affected by that specific action. The deletion of a link requires the deletion of several rows of the truth table. A natural choice would dictate that rows where the deleted input node has an activation equal to 0 are deleted. We can note however that such a choice is equivalent to give a semantic meaning to the symbol "0": that of "absence of signal". Consequently, to the symbol "1" corresponds the signal of "active input node". Semantics therefore implies a precise choice of the lines to be deleted.

A similar observation can be made in the case an input connection is added. We need to double the truth table: in the case where the additional node is zero (half the rows of the new table) we decide to keep the outputs of the original table: this corresponds to the situation where the incoming node is not active (so we keep the previous behavior, when the added node was not present). In the case the added node is active (the other half of the rows of the new table) decisions have to be made: in this situation the new element in fact exerts its influence. The choice in this work is that of randomizing the outputs of the original truth table, by keeping the original bias.

In summary, it should be kept in mind that the choices made in this way give semantic meaning to the value 0 and 1, thus making the symmetry between b and $1 - b$ no longer valid.

Following [15] we make use of a other type of mutation:

- gene duplication: an exact copy of an existing gene is created, including its input and output links: the truth table of the original node is duplicated on the new node. As consequence, the nodes to which the original node sends signals have an extra input link: thus, the mutation rules presented above apply to them.

  o   and divergence: the process of copy may not be perfect; in addition, the mutations described above may subsequently apply differently to the two copies of the same gene. The two genes then become increasingly different: this is what is known as genetic divergence [33–35].

Indeed, there is interest in observing what happens simply by adding a (slightly mutated) copy of an existing gene to an already functioning system [15]. In this work we follow this approach.

In order to simulate gene duplication followed by divergence, it is useful to first analyze in detail a single time step, investigating whether the two requirements of

i)   keeping the same attractors and (Requirement1)
ii) adding at least one new attractor (Requirement2)

are met under different choices for steps 4 and 6 of the algorithm depicted in Fig. 1.

As anticipated, this approach highlights characteristics that may be common to many kinds of fitness [15]. Thus, the observations made on the so shaped evolutionary process could acquire an interesting generality.

In the following we will present results concerning two cases of gene duplication and divergence:

a)   One input and one output connection of the duplicated node, chosen at random, are rewired at random. The Boolean function associated to the node is modified by changing a randomly chosen bit of its truth table (slightly changing in such a way the bias of the Boolean function). This move is called CH1 in the following.
b)   Each input and output connection are rewired at random with probability equal to 0.5, independently of the others. The Boolean function associated to the node is randomly changed, by keeping the same bias. This move is called CH2 in the following.

Finally, in genetic algorithms populations of individuals are typically used, with possible genetic exchange processes. However, it is not uncommon to follow single lineages, an approach included in the classic GA assuming low values of crossover probability [19, 36]. This modality allows a better comprehension of the effects of the evolutionary moves. In Fig. 1 is illustrated a high-level pseudo-code version of the used evolutionary process.

Algorithm 1 Aldana Evolution

1: $\mathbf{P} \leftarrow RBN$
2: **while** Termination condition **do**
3:    **n** $\leftarrow$ Randomly chosen node
4:    **n'** $\leftarrow$ Duplicate node of **n**
5:    **P'** $\leftarrow$ **P** with **n'** added
6:    Make_some_mutation(**P'**,**n'**)
7:    **next** $\leftarrow$ True
8:    **foreach ATTR** in List_of_attractor(**P**) **do**
9:     **if ATTR** not in List_of_attractor(**P'**) **then**
10:     **next** $\leftarrow$ False
11:   **if next** == True **then**
12:    **P** $\leftarrow$ **P'**
13: **end**

Fig. 1. A high-level pseudo-code version of the evolutionary approach used in this work. The functions present in the algorithm can be described as follows. **Make_some_mutation**: adds one or more mutations (chosen from those listed in the main text) to a gene in the RBN. **List_of_attractor**: returns the list of the attractors of the network given as input. Regarding the end condition of the while loop, we can opt among several choices: it can involve (i) a fixed number of iterations or (ii) a fixed number of successful mutations, or (iii) the maximum number of genes up to which the system can be expanded.

4 Results

The dynamical properties of random Boolean networks can be different in different instances, even if they are generated with the same parameters. Therefore, in order not to be misled by the idiosyncrasies of some specific exemplars, the attention is focused on the typical (or average) properties of a statistical ensemble of networks, all generated with certain parameter values. The average input connectivity per node, k, and the so-called bias b of the randomly generated functions (i.e., the fraction of 1's in their truth tables), are the most important parameters.

In this work we are interested in exploring the relationship between the dynamical regime of the network and its capability to evolve. For this purpose, we consider initial ensembles with different characteristics, classified according to their dynamic regime. In particular, we performed numerical experiments with different values of the number of connections per node and of the bias. Note that the critical value of the bias is $b = 0.5$ for networks with two connections per node, and $b \cong 0.21$ for networks with $k = 3$. The key question is whether critical networks can perform better than those with different dynamical regimes. This evaluation refers to a single iteration step, but it might provide useful information on the whole evolutionary process.

Some interesting results are summarized here below.

In Fig. 2a one can see the effect of changing the number of input connections per node, k, at a fixed bias ($b = 0.5$), and in Fig. 2b the effect of changing the bias by keeping the same average number of input connections per node ($k = 2$).

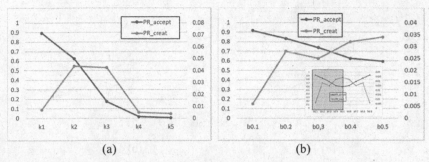

Fig. 2. The effect of changing connectivity or bias in RBN systems, in case of the CH1 move (quite similar results were obtained with the CH2 move). (a) Case $b = 0.5$, varying connectivity ($k \in [1, 5]$, k1, …, k5 in the figure). The fraction of cases where the attractor landscape of the transformed RBN contains all the original attractors is shown by a blue line, while the fraction of cases where it ALSO contains at least a new attractor is shown in grey, as a function of the network connectivity k. The scale on the left refers to the blue line, the one on the right to the grey line. (b) The fraction of cases where the attractor landscape of the transformed RBN contains all the original attractors is shown by a blue line, while the fraction of cases where it ALSO contains at least a new attractor is shown in grey, as a function of the bias (case with $k = 2$). The scale on the left refers to the blue line, the one on the right to the grey line. The insert shows the complete figure, up to bias $= 0.9$, assuming that a simple reflection of the data already acquired is correct enough. The simulations were carried out for networks with $N = 20$ and 10,000 network realizations.

One can see that the simultaneous satisfaction of Requirement1 and Requirement2 (i) is much higher when $k = 2$ or $k = 3$ than for larger or smaller values of the average degree of the nodes. One cannot tell from the observed values which one is larger. $k = 2$ corresponds in this case to dynamical criticality, while networks with $k = 3$ are mainly in the chaotic region. Note also that, when $k = 2$, in 60% of the cases all the old attractors are conserved, while this percentage falls to 20% for $k = 3$ (data not shown).

In Fig. 3b one can see the effect of changing the bias while keeping the number of connections per node constant. This graph shows that in the case $k = 2$ critical networks have the largest potential to acquire new attractors while preserving all the old ones. It is interesting to focus the attention to RBNs with $k = 3$, where all three regimes (ordered, disordered, chaotic) can be present. In these systems indeed the maximum is reached for bias values somewhat larger than the critical value 0.21 - Fig. 3a, CH1 move. The same happens in case of CH2 type of mutation (Fig. 3b): one can observe that the maximum capability to keep the old attractors and generate at least a new one is again reached for values of the connectivity bias which seem to fall in the chaotic region, slightly beyond the critical zone. In the chosen evolutionary process, therefore, the critical dynamic regime appears to be an important aspect - but other factors may also play a role.

An interesting observation concerns the direction in which evolution seems to proceed as it has been modeled so far. We measured the dynamical regime of the initial RBNs and that of their descendants, divided between those who meet the fitness constraints and those who do not. In both cases, typically we measured an increase in the Derrida coefficient, with a smaller increase for the accepted descendants. The addition of a gene (with the consequent increase in the input connectivity of some genes of the

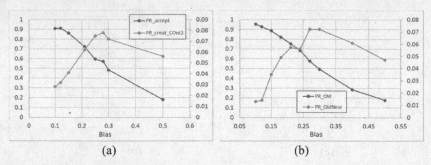

(a)   (b)

Fig. 3. The fraction of cases where the attractor landscape of the transformed RBN contains all the original attractors is shown by a blue line, while the fraction of cases where it ALSO contains at least a new attractor is shown in grey, as a function of the bias. The scale on the left refers to the blue line, the one on the right to the grey line (case with $k = 3$). In (a), by using the CH1 approach; in (b), by using the CH2 approach. The simulations were carried out for networks with $N = 20$, $k = 3$ and 1,000 network realizations.

RBN and the possible modification of the average bias of the system) therefore tends to amplify the dynamic disorder of the RBNs: the constraint of maintaining the attractors of the ancestors counteracts this trend, without completely eliminating it (see Fig. 4). Indeed, this observation is in agreement with Aldana's later works, such as [32].

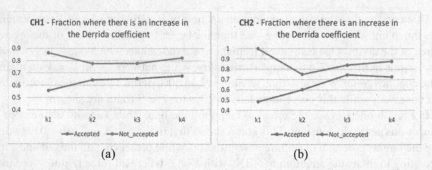

(a)   (b)

Fig. 4. The fraction of cases where there is an increase in the value of the Derrida coefficient with respect to the initial situation, for the accepted (blue line) and not accepted (orange line) descendants, in case of (a) CH1 mutation and (b) CH2 mutation. In both cases this fraction is always equal to or higher (often significantly higher) than the value 0.5: the majority of mutations therefore tend to shift the dynamic regime of the descendants towards conditions of greater disorder than the progenitors. The situation shown concerns RBNs with $N = 20$, $b = 0.5$, 10000 simulations.

5   Results

The principle of dynamical criticality is a very important one, and it therefore deserves a thorough investigation. The proposal to model gene duplication followed by divergence is a sound one, which we have subjected to extensive testing. The data obtained so far seem to indicate that there may be something else involved besides criticality, when the symmetry between the two Boolean values is broken. One might guess that this be related to the fact that, in our simulations, the value "1" means that a gene is expressed, while "0" denotes a silent gene. This choice leads to some slight differences in the way in which different changes are performed for $k = 3$, either in CH1 or in CH2. Although we had guessed that these effects were negligible, this might perhaps not be the case and it might lead to a situation where the maximum effect does not take place at the critical value.

An interesting alternative approach might be that of modifying the set of experiments in a way which imitates more closely the real biological processes of gene duplication followed by divergence. In this case, the fact that slightly chaotic nets outperform the critical ones might be attributed to the difference between the stylized model and biological reality. A closer description of biological phenomena might be obtained by introducing a more gradual way to change the network (e.g., by modifying only the outgoing links of the duplicated nodes in a way which should simulate the effect of a few mutations in its downstream DNA sequence) or by using only biologically plausible Boolean functions (for example, canalizing functions [37–39]).

A more comprehensive approach, which we plan to follow in the future, would be that of simulating the evolution of the whole network, for a large number of time steps, investigating the role of criticality. In this case the duplicated node might undergo different changes, but also the other nodes might do the same, although perhaps at a slower pace.

Moreover, the requirement of keeping all the old attractors might be too strong from a biological viewpoint, so one might consider relaxing it to some extent.

Funding. This research was funded by Università degli Studi di Modena e Reggio Emilia (FAR2019 project of the Department of Physics, Informatics and Mathematics).

References

1. Kauffman, S.A.: Metabolic stability and epigenesis in randomly constructed genetic nets. J. Theor. Biol. **22**(3), 437–467 (1969)
2. Kauffman, S.A.: The Origins of Order: Self-Organization and Selection in Evolution. Oxford University Press, Oxford (1993)
3. Kauffman, S.A.: At Home in the Universe: The Search for Laws of Self-Organization and Complexity. Oxford University Press, Oxford (1995)
4. Packard, N.H.: Adaptation toward the edge of chaos. Dyn. Patterns Complex Syst. **212**, 293–301 (1988)
5. Langton, C.: Life at the edge of chaos. Artificial life II. Santa Fe Institute Studies in the Science of Complexity (1992)
6. Crutchfield, J.P., Young, K.: Computation at the Onset of Chaos, Complexity, Entropy, and Physics of Information. Addison Wesley, New Jerseya (1990)
7. Serra, R., Villani, M., Semeria, A.: Genetic network models and statistical properties of gene expression data in knock-out experiments. J. Theor. Biol. **227**(1), 149–157 (2004)
8. Serra, R., Villani, M., Graudenzi, A., Kauffman, S.A.: Why a simple model of genetic regulatory networks describes the distribution of avalanches in gene expression data. J. Theor. Biol. **246**(3), 449–460 (2007)

9. Shmulevich, I., Kauffman, S.A., Aldana, M.: Eukaryotic cells are dynamically ordered or critical but not chaotic. Proc. Natl. Acad. Sci. **102**(38), 13439–13444 (2005)

10. Daniels, B.C., et al.: Criticality distinguishes the ensemble of biological regulatory networks. Phys. Rev. Lett. **121**(13), 138102 (2018)

11. Villani, M., La Rocca, L., Kauffman, S.A., Serra, R.: Dynamical criticality in gene regulatory networks. Complexity (2018)

12. Roli, A., Villani, M., Filisetti, A., Serra, R.: Dynamical criticality: overview and open questions. J. Syst. Sci. Complex. **31**(3), 647–663 (2018)

13. Munoz, M.A.: Colloquium: criticality and dynamical scaling in living systems. Rev. Mod. Phys. **90**(3), 031001 (2018)

14. Aldana, M., Coppersmith, S., Kadanoff, L.P.: Boolean dynamics with random couplings. In: Kaplan, E., Marsden, J.E., Sreenivasan, K.R. (eds.) Perspectives and Problems in Nolinear Science, pp. 23–89. Springer, New York (2003). https://doi.org/10.1007/978-0-387-217 89-5_2

15. Aldana, M., Balleza, E., Kauffman, S., Resendiz, O.: Robustness and evolvability in genetic regulatory networks. J. Theor. Biol. **245**(3), 433–448 (2007)

16. Benedettini, S., et al.: Dynamical regimes and learning properties of evolved Boolean networks. Neurocomputing **99**, 111–123 (2013)

17. Szejka, A., Drossel, B.: Evolution of canalizing Boolean networks. Eur. Phys. J. B **56**(4), 373–380 (2007)

18. Mihaljev, T., Drossel, B.: Evolution of a population of random Boolean networks. Eur. Phys. J. B **67**(2), 259–267 (2009)

19. Braccini, M., Roli, A., Villani, M., Serra, R.: Automatic design of Boolean networks for cell differentiation. In: Rossi, F., Piotto, S., Concilio, S. (eds.) WIVACE 2016. CCIS, vol. 708, pp. 91–102. Springer, Cham (2017). https://doi.org/10.1007/978-3-319-57711-1_8

20. Liu, M., Bassler, K.E.: Emergent criticality from coevolution in random Boolean networks. Phys. Rev. E **74**(4), 041910 (2006)

21. Magrì, S., Villani, M., Roli, A., Serra, R.: Evolving critical Boolean networks. In: Cagnoni, S., Mordonini, M., Pecori, R., Roli, A., Villani, M. (eds.) WIVACE 2018. CCIS, vol. 900, pp. 17–29. Springer, Cham (2019). https://doi.org/10.1007/978-3-030-21733-4_2

22. Villani, M., Magrì, S., Roli, A., & Serra, R.: Selecting for positive responses to knock outs in Boolean networks. In: Cicirelli, F., Guerrieri, A., Pizzuti, C., Socievole, A., Spezzano, G., Vinci, A. (eds.) Italian Workshop on Artificial Life and Evolutionary Computation, pp. 7–16. Springer, Cham (2019). https://doi.org/10.1007/978-3-030-45016-8_2

23. Derrida, B., Pomeau, Y.: Random networks of automata: a simple annealed approximation. EPL (Europhys. Lett.) **1**(2), 45 (1986)

24. Langton, C.G.: Computation at the edge of chaos: phase transitions and emergent computation. Physica D **42**(1–3), 12–37 (1990)

25. Bailly, F., Longo, G.: Extended critical situations: the physical singularity of life phenomena. J. Biol. Syst. **16**(02), 309–336 (2008)

26. Derrida, B., Flyvbjerg, H.: The random map model: a disordered model with deterministic dynamics. J. de Physique **48**(6), 971–978 (1987)

27. Bastolla, U., Parisi, G.: The modular structure of Kauffman networks. Physica D **115**(3–4), 219–233 (1998)

28. Bastolla, U., Parisi, G.: Relevant elements, magnetization and dynamical properties in Kauffman networks: a numerical study. Physica D **115**(3–4), 203–218 (1998)

29. Holland, J.H.: Adaptation in Natural and Artificial Systems: An Introductory Analysis with Applications to Biology, Control, and Artificial Intelligence. MIT Press (1992)

30. Roli, A., Benedettini, S., Birattari, M., Pinciroli, C., Serra, R., Villani, M.: Robustness, evolvability and complexity in Boolean network robots. In: Proceedings of the ECCS2011, Vienna, Austria, pp. 12–16 (2011)

31. Roli, A., Villani, M., Serra, R., Benedettini, S., Pinciroli, C., Birattari, M.: Dynamical properties of artificially evolved Boolean network robots. In: Gavanelli, M., Lamma, E., Riguzzi, F. (eds.) AI*IA 2015. LNCS (LNAI), vol. 9336, pp. 45–57. Springer, Cham (2015). https://doi.org/10.1007/978-3-319-24309-2_4

32. Torres-Sosa, C., Huang, S., Aldana, M.: Criticality is an emergent property of genetic networks that exhibit evolvability. PLoS Comput. Biol. **8**(9) (2012)

33. Lynch, M., Conery, J.S.: The evolutionary fate and consequences of duplicate genes. Science **290**(5494), 1151–1155 (2000)

34. Lynch, M.: Gene duplication and evolution. Science **297**(5583), 945–947 (2002)

35. Zhang, J.: Evolution by gene duplication: an update. Trends Ecol. Evol. **18**(6), 292–298 (2003)

36. Blum, C., Roli, A.: Metaheuristics in combinatorial optimization: overview and conceptual comparison. ACM Comput. Surv. (CSUR) **35**(3), 268–308 (2003)

37. Harris, S.E., Sawhill, B.K., Wuensche, A., Kauffman, S.: A model of transcriptional regulatory networks based on biases in the observed regulation rules. Complexity **7**(4), 23–40 (2002)

38. Just, W., Shmulevich, I., Konvalina, J.: The number and probability of canalizing functions. Physica D **197**(3–4), 211–221 (2004)

39. Karlsson, F., Hörnquist, M.: Order or chaos in Boolean gene networks depends on the mean fraction of canalizing functions. Physica A **384**(2), 747–757 (2007)

Effective Resistance Based Weight Thresholding for Community Detection

Clara Pizzuti and Annalisa Socievole[(✉)]

National Research Council of Italy (CNR), Institute for High Performance
Computing and Networking (ICAR), Via Pietro Bucci, 8-9C, 87036 Rende, CS, Italy
{clara.pizzuti,annalisa.socievole}@icar.cnr.it

Abstract. Weight thresholding is often used as a sparsification procedure when graphs are too dense and reducing the number of edges is necessary to apply standard graph-theoretical methods. This work presents a
proof-of-principle of a new evolutionary method based on Genetic Algorithms detecting communities in weighted networks by exploiting the
concepts of effective resistance and weight thresholding. Given an input
weighted graph, the algorithm considers its equivalent electric network
where the edge weights are recomputed taking also into account the
effective resistance, whose square root has shown to be a Euclidean metric. The method then generates a weighted sparse graph by maintaining
for each node only the neighbors having weights below a given distance
threshold. In such a way, only the neighbor nodes with low effective
resistances and thus, highly similar according to this metric, are considered. Experiments on synthetically generated networks show that our
approach is effective when compared to other benchmarks.

Keywords: Community detection · Effective resistance · Genetic
algorithms · Graph sparsification

1 Introduction

Many real-world systems in nature can be modeled as networks with interacting
nodes, where *nodes* correspond to the objects of the system and the interaction
are modeled with *edges* representing the relationship between the objects. Many
researchers have devoted the study of complex networks focusing on their *community* structure [9,11], i.e. the division of a network in groups of nodes having
dense intra-connections and sparse inter-connections. In several fields including
computer science, telecommunications, biology and physics, the presence of communities indicate a particular organization of the system in groups whose study
has many interesting applications [15]. As an example, uncovering the community structure of a collaboration network can give insights for partnership
recommendations, outlier detection, targeted advertisements, and so on. The
majority of the existing community detection algorithms are based on consensus
clustering, spectral methods, statistical inference, optimization techniques or are
dynamics-based [9].

J. J. Schneider et al. (Eds.): WIVACE 2021, CCIS 1722, pp. 14–23, 2022.
https://doi.org/10.1007/978-3-031-23929-8_2

An important feature of real-world networks is that nodes are usually connected through many weighted edges resulting in highly dense graphs [3]. In such cases (e.g. financial or brain networks), edge cut strategies can help in making those graphs usable since they aim at removing weak or not important connections for graph-theoretical analysis of dense weighted structures. Different strategies of edge *pruning* or *graph sparsification* have been proposed so far [18,22,23]. The goal of graph sparsification is to build an approximate graph H of a given dense graph G on the same set of vertices. If H is similar to G according to some appropriate metric, then H can be used as a proxy for G in computations without introducing too much error. *Weight thresholding* is the most popular approach of graph sparsification consisting in cutting all the edges that are below a given threshold [26]. By analyzing several synthetic and real-world weighted networks, it is shown how this procedure does not alter the community structure even if most of the edges are removed. In another work [22], each edge of G is included in the sparsifier H with a probability proportional to their *effective resistance* proving that such a sampling of the input graph G yields to a good sparsifier. The effective resistance [10,13] is increasingly gaining attention in this research area due to its appealing properties. Viewing an undirected graph as an electrical network in which each edge of the graph is replaced by a unit resistance, the effective resistance of an edge represents the probability that the edge appears in a random spanning tree of G [8] and is intimately connected to the *length* of random walks on the graph. It was also proven to be proportional to the commute time between its endpoints [5]. Moreover, [13] showed that the square root of the effective resistance between any couple of nodes (i, j) is a Euclidean metric, measuring the distance between nodes i and j. Thus, the computation of the effective resistance for each edge of G provides a distance matrix between each couple of nodes of G.

In this work, we focus on community detection on weighted networks by both considering weight thresholding as sparsification procedure of the input graph and effective resistance as metric to further weight edges. Given a weighted undirected network $G = (V, E)$, we propose a community detection algorithm based on *Genetic Algorithms (GAs)* [12], namely *WomeGAnet*. The main idea is to find communities on the weighted graph $G' = (V, W)$ obtained from G by computing the effective resistance of all the node pairs of G and weighting the resulting effective resistance edge value with the initial edge weight. G' has the same set V of nodes of G, while the set W of edges consists of the re-weighted edges between any couple of nodes (i, j) of V. Since the resulting adjacency matrix Ω corresponding to G' is a full matrix, we also apply weight thresholding [26] by considering for each node i only a fixed number nn nearest neighbors to obtain a sparse weighted adjacency matrix $\widetilde{\Omega}$. We therefore run the GA over $\widetilde{\Omega}$ by evolving a population of individuals and minimizing *modularity* [11] as objective function. To validate our method, we compare it both with a baseline genetic algorithm named *GA-mod*, optimizing modularity and running on the original graph G, and the two benchmarks *Louvain* [4] and *Infomap* [19]. By examining our method on a pool of synthetic networks, we show that *WomeGAnet* is competitive when network communities are separated by few inter-community links and even better than the contestant methods when the number inter-community links increases.

The paper is organized as follows. The next section describes the most relevant research works in this area. Section 3 recalls the concept of effective resistance. Section 4 defines the problem we tackle and describes the proposed community detection algorithm. Section 5 describes the dataset used and the experiments performed to validate *WomeGAnet*. Finally, Sect. 6 concludes the paper.

2   Related Work

In the literature of distance measures, many definitions have been proposed so far. Such measures are fundamental in graph clustering tasks, for example, since they are able to provide a criterion for grouping nodes according to their proximity/similarity. The most common distances between two nodes include Shortest Path [7], Resistance [13], Logarithmic Walk, Forest, Heat, Communicability and many others [1, 6]. *Effective resistance* was proposed by Klein and Randic [13] as another distance measure with interesting properties. First, it has been shown to be an Euclidean metric. Moreover, the effective resistance precisely characterizes the *commute time* [5], the expected length of a random walk from a node to another and back. This is why the effective resistance is also named *commute time distance*. In [20], the relationship between the Laplacian matrix of the graph and the commute time is investigated showing that the commute time distance can be computed through the *pseudoinverse* of the Laplacian, which is a *kernel* (a well-defined similarity measure between the nodes).

The commute time kernel was exploited for clustering the nodes of a weighted undirected graph [27]. The proposed method first computes the sigmoid commute time kernel matrix from the adjacency matrix of the graph, thus obtaining a similarity measure between nodes. Then, it clusters the nodes by exploiting a kernel-based k-means or fuzzy k-means on the obtained kernel matrix. By applying the method to a document clustering problem involving the newsgroups database, the results highlight its superiority when compared to classical k-means, as well as spectral clustering. In a later study [28], the same two-step procedure was compared to a selection of graph clustering and community detection algorithms on three real-world databases. The results of such analysis show that the joint use of the commute time kernel matrix and kernel clustering result quite effective. In another work [21], the effectiveness of the distance-based algorithms is assessed through a comparison between six different distance measures transformed into kernels and tested on kernel k-means and a weighted version of it is made. A comparison with *Louvain* benchmark shows the effectiveness of the distance-based algorithms.

The role of network topology on the efficiency of proximity measures for community detection is analyzed in [2]. The results of such analysis show that it is possible to find measures that behave well for most cases. Finally, the impact of graph sparsification through weight thresholding on the community detected is investigated in [26]. By introducing the MASS, a measure estimating the variation of the spectral properties of the graph when edges are removed, it is shown that the group structure of real weighted networks is very robust under weight thresholding, in contrast to other features. This is shown to be due to the peculiar correlation between weights and degree commonly observed on real networks, for which large weights commonly characterize the links of high-degree nodes. A major advantage of MASS is its numerical stability and computational efficiency.

3 Preliminaries

In this section, we recall the concept of *effective resistance* [13]. This measure is a graph theory concept originated from resistive electrical networks: if each edge of a connected graph is characterized by a resistor, then the effective resistance between couples of nodes is a graphical distance capturing the global properties of the network, i.e., both direct and indirect connections between two nodes. Intuitively, the effective resistance measures the closeness between nodes when multiple paths are available. The existence of several paths between two nodes reduces the distance: two nodes separated by many paths are closer than two nodes separated by fewer paths.

More formally, given an undirected and connected graph $G = (V, E)$, we can associate to G an *equivalent electric network* where each edge $(i, j) \in E$ has a weight w_{ij} representing its conductance, i.e. the inverse of the electrical resistance ω_{ij} of a resistor, so that $w_{ij} = \frac{1}{w_{ij}}$ *ohm* [13].

For any edge of the graph G, a distance function can be defined as follows [13,25]. Let A be the adjacency matrix of G with elements $a_{ij} = 1$ if there is an edge between i and j. Given $\Delta = diag(d_i)$ the $N \times N$ diagonal degree matrix, where $d_i = \sum_{j=1}^{N} a_{ij}$, the Laplacian matrix L of the graph G is defined as the $N \times N$ symmetric matrix $L = \Delta - A$, with elements

$$l_{ij} = \begin{cases} d_i & \text{if } i = j \\ -1 & \text{if the edge } (i, j) \in E \\ 0 & \text{otherwise} \end{cases} \tag{1}$$

The effective resistance ω_{ij} between two nodes i and j is defined as the voltage developed between i and j when a unit electric current flows from i to j. To include all the effective resistances ω_{ij} of G, we use the $N \times N$ matrix Ω. Ω can be defined [8,13,24] as

$$\Omega = \zeta u^T + u \zeta^T - 2L^+ \tag{2}$$

where the vector

$$\zeta = (L_{11}^+, L_{22}^+, \ldots, L_{NN}^+) \tag{3}$$

contains the diagonal elements of the Moore-Penrose *pseudoinverse* matrix L^+ of the weighted Laplacian matrix $\tilde{L}$ of G. More specifically, the effective resistance between two nodes x and y equals

$$\omega_{xy} = (e_x - e_y)^T L^+ (e_x - e_y) = l_{xx}^+ + l_{yy}^+ - 2l_{xy}^+ \tag{4}$$

where e_k is the basic vector with the m-th component equal to $(e_k)_m = \delta_{mk}$ and δ_{mk} is the Kronecker-delta: $\delta_{mk} = 1$ if $m = k$, otherwise $\delta_{mk} = 0$.

The effective resistance is widely applied in several theoretic and also practical problems. Several interesting features characterize this measure. The shortest path distance in a graph, for example, upperbounds ω_{ij} [24]. Moreover, the commute time distance C_{ij} between two nodes i and j, i.e. the expected number of steps needed during a random walk from i to j, is $C_{ij} = u^T \tilde{A}u \, \omega_{ij}$, where u is the all one vector and $u^T \tilde{A}u$ is the double of the sum of all the edge weights in the weighted adjacency matrix $\tilde{A}$ [5]. In addition, the square root $\sqrt{\omega}_{ij}$ of the effective resistance is a Euclidean metric [13]. More precisely, the effective resistance matrix Ω is a distance matrix, in which for any triple of non-negative elements, $\omega_{ii} = 0$, and the triangle inequality, $\omega_{ij} \leq \omega_{ik} + \omega_{kj}$ is satisfied.

4   The *WomeGAnet* Method

We now describe the proposed community detection method, namely *WomeGAnet*, by first introducing the problem we tackle and then providing a description of the algorithm.

Problem Definition. Given a weighted graph G, let Ω be its resistance matrix, $\tilde{A}$ its weighted adjacency matrix, nn the number of nearest neighbors to consider, and $\tilde{\Omega}$ the matrix obtained from Ω with elements

$$\tilde{\omega}_{xy} = \begin{cases} \tilde{a}_{xy}\omega_{xy} & \text{if } y \text{ is among the } nn \\ & \quad \text{nearest neighbors of } x \\ 0 & \text{otherwise} \end{cases} \tag{5}$$

The community detection problem to solve is: *find a partition* $C = \{C_1, ..., C_k\}$ *of the nodes of* G *such that the weighted modularity* Q *of* C *is maximized*, where

$$Q = \frac{1}{2m} \sum_{ij} \left(\tilde{\omega}_{ij} - \frac{k_i k_j}{2m} \right) \delta(C_i, C_j) \tag{6}$$

and m is the sum of the edge weights, k_i and k_j are the weighted degrees of nodes i and j respectively, and δ is the Kronecker function which yields one if i and j are in the same community, i.e. $C_i = C_j$, zero otherwise. Modularity expresses the quality of a partition by computing the expected number of edges within the communities of a random graph with the same degree distribution. We point out that our aim is two-fold: (1) find a partition where the intra-cluster weighted modularity is high and (2) the intra-community similarity between nodes is high (i.e. the overall distance between the nodes of the community is low).

In this paper, we use a Genetic Algorithm (GA) [12] as an effective optimization technique [17] solving the above community detection problem. The algorithm, namely *WomeGAnet*, initially creates a population of individuals that are randomly generated, where each individual represents a possible solution of the problem and hence, a network partition in communities. Subsequently, the method evolves the population of the individuals through the genetic operators of variation and selection by optimizing the value of the objective function (i.e. weighted modularity) while exploring the search space.

As genetic representation, *WomeGAnet* adopts the *locus-based* [16] for which the individual/chromosome is encoded through a string of n genes, where a gene is a node. When a value i from 1 to N is assigned to the k-th gene, it means that nodes k and i are connected. Then, a decoding step identifies all the connected components of the graph which correspond to the communities. The crossover operator used by the method is the *uniform crossover*. It generates an offspring where each gene is chosen from either parent with equal probability. Finally, the mutation operator randomly assigns the value of a i-th gene to one of its neighbors.

WomeGAnet receives in input the graph $G = (V, E)$, its adjacency weighted matrix $\widetilde{A}$, the number of nearest neighbors nn to consider and performs the following steps:

1. compute the Laplacian L of G;
2. compute the Moore-Penrose *pseudoinverse* matrix L^+ of L;
3. compute the effective resistance matrix Ω from L^+ as $\omega_{xy} = l^+_{xx} + l^+_{yy} - 2l^+_{xy}$;
4. make Ω a distance matrix by substituting each element as $\omega_{xy} = \sqrt{\omega_{xy}}$;
5. normalize the elements of Ω and make Ω a symmetric matrix $\Omega = 0.5[\frac{\Omega}{max(\Omega)} + (\frac{\Omega}{max(\Omega)})^T]$;
6. select for each node x only the nn entries ω_{xy} from Ω having the minimum distance value, i.e. the connections with the nodes y having the highest similarity values with x;
7. generate the sparse similarity weighted matrix $\widetilde{\Omega}$ including only the selected entries ω_{xy} and weighting them as $\widetilde{\omega}_{xy} = \widetilde{a}_{xy}\omega_{xy}$;
8. run the Genetic Algorithm on $\widetilde{\Omega}$ for a number of iterations by using modularity as fitness function to maximize, uniform crossover and neighbor mutation as variation operators;
9. obtain the partition $C = \{C_1, \ldots, C_k\}$ corresponding to the solution with the highest fitness value.

5   Experimental Evaluation

To test the proposed method, we carried out simulations using Matlab 2020a and the Global Optimization Toolbox. We generated a set of synthetic 128-nodes networks with realistic community structures by exploiting the Lancichinetti-Fortunato-Radicchi (LFR) benchmark [14]. The parameters used for producing the networks are shown in Table 1. In particular, for controlling the community structure of the network, we set the *mixing parameter* μ. Low values of μ generate communities with many inter-community and few inter-community links. Since all the edges have weight 1, we assigned a random weight between 0 and 1 to each of them. As genetic parameters for *WomeGAnet*, we set maximum number of generations 100, population size 700, crossover fraction 0.9 and mutation rate 0.1. Finally, for setting the number of nearest neighbors nn to consider for each node, we evaluated the MASS [26] measure as a function of the fraction of edges removed. As described in Sect. 2, this simple measure allows to estimate the variation of the spectral properties of the graph under weight thresholding. Since the spectral properties of a graph are theoretically related to its underlying community structure, the MASS can be thus exploited to establish the threshold of edges to disconnect above which the community structure of the network would be altered. For each network, we computed the MASS for different fractions of edge removals ranging from 0.1 to 0.9. Each fraction of edge cuts represents a percentage of edges of the network having the highest weights. For each fraction of edge cuts considered, we obtained a MASS value of 1. In other words, the networks considered are robust both when few or many edge fractions are removed. As such, we can choose any number of nearest neighbors to disconnect being sure to not change the community structure and hence, to not impact the results of the algorithm. Note that the number of nearest neighbors is chosen in a range where the minimum value is 1 (only one neighbor) and the maximum number of neighbors which can be generated by the LFR benchmark. In our case, this value is 9. We thus chose $nn = 4$, which is a reasonable value considering that the node average degree is 8.

We first compared *WomeGAnet* to a baseline GA-based algorithm, namely *GA-mod*, characterized by the same locus-based representation, initialization, crossover and mutation operators and parameter settings, and optimizing modularity as objective function. Moreover, during the population initialization, *GA-mod* connects a node with one of its nn nearest neighbors, instead of a random neighbor. We also compared *WomeGAnet* to the two well-known methods *Louvain* [4] and *Infomap* [19]. The first is based on a greedy modularity optimization approach. Specifically, *Louvain* first identifies small communities by locally optimizing modularity. Then, it builds a new network whose nodes are the communities previously found, and these steps are repeated until a hierarchy of high-modularity communities is obtained. *Infomap*, on the contrary, exploits the principles of information theory by defining the community detection problem as the problem of finding a description of minimum information of a random walk on the graph. The method maximizes the Minimum Description Length as objective function by quickly providing an approximation of the optimal solution.

Table 2 shows the results for mixing parameters $\mu = 0.1, 0.2, ..., 0.5$. We used the *Normalized Mutual Information* (NMI) to measure the similarity between the obtained solutions and the ground-truth. Each value has been averaged over 30 runs of the method. For communities with clear structure ($\mu = 0.1$ and $\mu = 0.2$), the two genetic algorithms and *Infomap* match the ground truth correctly identifying the underlying communities. *Louvain*, on the contrary, achieves only 0.7894 and 0.7645 for $\mu = 0.1$ and $\mu = 0.2$, respectively. When μ increases (μ from 0.3 to 0.5) and the communities become less clear, *WomeGAnet* shows its superiority by outperforming all the other methods. Overall, the NMI decreases achieving

Table 1. Parameter settings for the LFR-128 synthetic networks.

Parameter	Value
Number of nodes	128
Node average degree	8
Node maximal degree	9
Exponent for power law creating degree sequence	2
Exponent for power law creating community sizes	1
Mixing parameter μ	[0.1; 0.6]
Maximal community size	40
Minimal community size	20
Average density	0.062

Table 2. NMI and standard deviation results for 30 runs.

	WomeGAnet	GA-mod	Louvain	Infomap
$\mu = 0.1$	1 (0)	1 (0)	0.7894 (0)	1 (0)
$\mu = 0.2$	1 (0)	1 (0)	0.7645 (0)	1 (0)
$\mu = 0.3$	0.6562 (0.0917)	0.571 (0.1022)	0.4521 (0)	0.491 (0.062)
$\mu = 0.4$	0.4249 (0.0922)	0.3392 (0.0541)	0.3964 (0)	0.325 (0.0271)
$\mu = 0.5$	0.2692 (0.0198)	0.1281 (0.0917)	0.2055 (0)	0.1228 (0.0275)

Table 3. Number of communities results for one single run.

	GT	WomeGAnet	GA-mod	Louvain	Infomap
$\mu = 0.1$	5	5	5	5	5
$\mu = 0.2$	5	5	5	3	5
$\mu = 0.3$	4	4	7	7	9
$\mu = 0.4$	4	4	7	7	6
$\mu = 0.5$	3	3	7	7	9

low values for $\mu = 0.5$. Looking at the communities found for a single run of the method (Table 3), *WomeGAnet* is always able to match the ground truth (GT).

6   Conclusions

We proposed *WomeGAnet*, a new method based on genetic algorithms for dividing the nodes of undirected and connected weighted graphs in communities. *WomeGAnet* has been designed by exploiting the *effective resistance* as distance measure between nodes and the concept of *weight thresholding* as graph sparsification procedure. More specifically, we considered the input graph as an electric circuit and computed for each couple of connected nodes the effective resistance. We then exploited this distance for computing the similarity between nodes and generated a weighted sparse graph by maintaining for each node only the neighbors having weights below a given distance threshold. We found that for sparsifying the considered graphs, the number of nearest neighbors chosen does not alter the community structure.

The experiments carried out on synthetically generated weighted networks show that *WomeGAnet* is effective since it clearly outperforms both a baseline GA-based algorithm running on the original adjacency matrix of the graph, and the benchmarks *Louvain* and *Infomap*. We point out that we have presented a proof-of-principle of *WomeGAnet*, which, however, deserves further investigations, ad hoc analyses and simulations in different scenarios, both synthetic with different generation parameters and real-world. In addition in the weight thresholding process, the choice of an arbitrary number of nearest neighbors has been driven by the robust structure of the networks considered. For future work, more study will be also devoted to the assessment of *WomeGAnet* performance when the MASS varies.

References

1. Avrachenkov, K., Chebotarev, P., Rubanov, D.: Kernels on graphs as proximity measures. In: Bonato, A., Chung Graham, F., Prałat, P. (eds.) WAW 2017. LNCS, vol. 10519, pp. 27–41. Springer, Cham (2017). https://doi.org/10.1007/978-3-319-67810-8_3
2. Aynulin, R.: Impact of network topology on efficiency of proximity measures for community detection. In: Cherifi, H., Gaito, S., Mendes, J.F., Moro, E., Rocha, L.M. (eds.) COMPLEX NETWORKS 2019. SCI, vol. 881, pp. 188–197. Springer, Cham (2020). https://doi.org/10.1007/978-3-030-36687-2_16
3. Barrat, A., Barthelemy, M., Pastor-Satorras, R., Vespignan, A.: The architecture of complex weighted networks. Proc. Nat. Acad. Sci., **101**, 3747 (2004)
4. Blondel, V.D., Guillaume, J.L., Lambiotte, R., Lefebvre, E.: Fast unfolding of communities in large networks. J. Stat. Mech: Theory Exp. **2008**(10), P10008 (2008)
5. Chandra, A.K., Raghavan, P., Ruzzo, W.L., Smolensky, R., Tiwari, P.: The electrical resistance of a graph captures its commute and cover times. Comput. Complex. **6**(4), 312–340 (1996)
6. Deza, M.M., Deza, E.: Encyclopedia of distances. In: Encyclopedia of Distances, pp. 1–583. Springer, Heidelberg (2009). https://doi.org/10.1007/978-3-642-00234-2_1
7. Dijkstra, E.W., et al.: A note on two problems in connexion with graphs. Numer. Math. **1**(1), 269–271 (1959)
8. Doyle, P., Snell, J.: Random walks and electric networks. Mathematical Association of America, Washington, D.C. (1989)
9. Fortunato, S., Hric, D.: Community detection in networks: a user guide. Phys. Rep. **659**, 1–44 (2016)
10. Ghosh, A., Boyd, S., Saberi, A.: Minimizing effective resistance of a graph. SIAM Rev. **50**(1), 37–66 (2008)

11. Girvan, M., Newman, M.E.: Community structure in social and biological networks. Proc. Natl. Acad. Sci. **99**(12), 7821–7826 (2002)
12. Goldberg, D.E.: Genetic Algorithms in Search, Optimization, and Machine Learning (1989)
13. Klein, D.J., Randić, M.: Resistance distance. J. Math. Chem. **12**(1), 81–95 (1993)
14. Lancichinetti, A., Fortunato, S., Radicchi, F.: Benchmark graphs for testing community detection algorithms. Phys. Rev. E **78**(4), 046110 (2008)
15. Newman, M.E.: The structure and function of complex networks. SIAM Rev. **45**(2), 167–256 (2003)
16. Park, Y., Song, M.: A genetic algorithm for clustering problems. In: Proceedings of the Third Annual Conference on Genetic Programming, vol. 1998, pp. 568–575 (1998)
17. Pizzuti, C.: Evolutionary computation for community detection in networks: a review. IEEE Trans. Evol. Comput. **22**(3), 464–483 (2018)
18. Radicchi, F., Ramasco, J.J., Fortunato, S.: Information filtering in complex weighted networks. Phys. Rev. E **E83**, 046101 (2011)
19. Rosvall, M., Bergstrom, C.T.: Maps of random walks on complex networks reveal community structure. Proc. Natl. Acad. Sci. **105**(4), 1118–1123 (2008)
20. Saerens, M., Fouss, F., Yen, L., Dupont, P.: The principal components analysis of a graph, and its relationships to spectral clustering. In: Boulicaut, J.-F., Esposito, F., Giannotti, F., Pedreschi, D. (eds.) ECML 2004. LNCS (LNAI), vol. 3201, pp. 371–383. Springer, Heidelberg (2004). https://doi.org/10.1007/978-3-540-30115-8_35
21. Sommer, F., Fouss, F., Saerens, M.: Comparison of graph node distances on clustering tasks. In: Villa, A.E.P., Masulli, P., Pons Rivero, A.J. (eds.) ICANN 2016. LNCS, vol. 9886, pp. 192–201. Springer, Cham (2016). https://doi.org/10.1007/978-3-319-44778-0_23
22. Spielman, D.A., Srivastava, N.: Graph sparsification by effective resistances. Siam J. Comput. (40), 1913 (1996)
23. Tumminello, M., Aste, T., Matteo, T.D., Mantegna, R.N.: A tool for filtering information in complex systems. Proc. Nat. Acad. Sci. **102**, 10421 (2005)
24. Van Mieghem, P.: Graph Spectra for Complex Networks. Cambridge University Press, Cambridge (2010)
25. Van Mieghem, P., Devriendt, K., Cetinay, H.: Pseudoinverse of the Laplacian and best spreader node in a network. Phys. Rev. E **96**(3), 032311 (2017)
26. Yan, X., Jeub, L.G.S., Flammini, A., Radicchi, F., Fortunato, S.: Weight thresholding on complex networks. Phys. Rev. E **E98**, 042304 (2018)
27. Yen, L., Fouss, F., Decaestecker, C., Francq, P., Saerens, M.: Graph nodes clustering based on the commute-time kernel. In: Zhou, Z.-H., Li, H., Yang, Q. (eds.) PAKDD 2007. LNCS (LNAI), vol. 4426, pp. 1037–1045. Springer, Heidelberg (2007). https://doi.org/10.1007/978-3-540-71701-0_117
28. Yen, L., Fouss, F., Decaestecker, C., Francq, P., Saerens, M.: Graph nodes clustering with the sigmoid commute-time kernel: a comparative study. Data Knowl. Eng. **68**(3), 338–361 (2009)

An Oracle for the Optimization of Underconstrained Compositions of Neural Networks - The Tick Hazard Use Case

Gregory Gygax[1], Nils Ratnaweera[2], Werner Tischhauser[2], Theo H. M. Smits[2], Patrick Laube[2], and Thomas Ott[1(✉)]

[1] Institute of Computational Life Sciences, Zurich University of Applied Sciences ZHAW, Wädenswil, Switzerland
ottt@zhaw.ch
[2] Institute of Natural Resource Sciences, Zurich University of Applied Sciences ZHAW, Wädenswil, Switzerland

Abstract. Modeling complex real-world variables, such as tick hazard, can face the problem that no measurements are available for the target variable itself. Models developed on the basis of indirect measurements may then lead to an underconstrained problem for which the output cannot be reliably validated. To address such a problem in the tick hazard use case, we propose a novel oracle approach. The goal of the oracle is to generate test scenarios that can be used to test the validity and robustness of our tick hazard model. We report on preliminary results that support the potential of both the model and the oracle approach.

1 Introduction

1.1 Problem Motivation

This contribution addresses a problem that has arisen in an ongoing research project whose goal is to create a spatio-temporal model of tick hazard in Switzerland based on reports from volunteers. These volunteers are ordinary citizens who use a smartphone app to report tick sightings and bites. Hence, the data we obtain from this source are a sample from a spatio-temporal distribution of human-tick encounters. Arguably, this distribution is a product of both human and tick activity, or their respective spatial distributions. The challenge is therefore to design a model that can be trained using this data sample, yet allowing to compute human and tick activity independently. We choose a model that combines two neural networks for both activity levels. However, it comes with a difficulty: since we have no ground truth for either human or tick activity, we cannot use the data to validate our model at these levels. Following ecological model validation procedures [1], the model outputs could be validated by an expert panel. This expert validation is ongoing and the results will be published elsewhere. In this work, we will present another validation scheme (the 'oracle' approach) that focuses on the validity of our data pipeline and model training setup in general. The aim of the scheme is

J. J. Schneider et al. (Eds.): WIVACE 2021, CCIS 1722, pp. 24–31, 2022.
https://doi.org/10.1007/978-3-031-23929-8_3

to show that our model is methodologically suited to the problem. Ultimately, we want to test our approach for robustness when dealing with increasingly noisy and uncertain data. Understanding this robustness enables us to provide a measure of confidence in our model. Comparable approaches exist in the field of computational vision, where synthetic datasets are used for training and the validation of image recognition tasks, e.g. in [3].

1.2   General Problem and Method Statement: The Oracle

Before we focus on the specific tick hazard use case, we first outline the problem and our approach in general terms. We face the problem of training a composition of neural networks for which the outputs of the individual networks are of interest but cannot be directly validated. As depicted in Fig. 1, the composition consists of two networks (NN1 and NN2). Labels for the combined output $O_3 = O_1 \times O_2$ are available and hence O_3 can be validated. We are, however, interested in the outputs of the individual networks O_1 and O_2 for which no labels are available. The problem is underconstrained as, for instance, O_1 and O_2 could be switched without changing the product O_3. How can we faithfully test that, or under what conditions, the composite model produces useful, i.e., realistic, intermediate outputs O_1 and O_2? And how can we optimize the hyperparameters of the composite model to find an adequate complexity?

We propose to use an artificial oracle - a term coined by software engineers for testing mechanisms [2] - to mimic real world data. The oracle is able to easily simulate world scenarios (pseudo-ground truth) that are used to test the accuracy and robustness of the composite model. The idea comprises the following aspects:

1. Accuracy is measured by comparing the outputs of the model and the oracle at all levels (O_1, O_2, O_3).
2. Robustness is essentially measured by varying the oracle's output ('noisy oracle') and studying the decay of the accuracy.

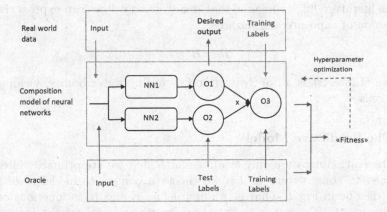

Fig. 1. Schematic sketch of the problem, model and oracle scheme (see text).

3. Based on a notion of fitness (objective function) as a combination of accuracy and robustness, the oracle enables hyperparameter optimization, e.g., the optimization of the number of layers or hidden nodes in the neural networks. This possibility is crucial since the actual complexity and the interactions of the real world input factors are unknown. If we can show that hyperparameter optimization leads to a model that is able to predict the outputs of an oracle even with high complexity, we can be confident that the approach will perform reasonably on real data.

2   The Tick Hazard Use Case

2.1   Hazard, Exposure and Risk

The goal is to develop a spatial model for predicting tick activity (*tick hazard* H) dependent on environmental parameters (such as type of land use [4], climate data, elevation, vegetation, etc.), see [5] for a comparable study. The only labels we have is observational data from volunteers. In particular, we are given data of around 50'000 reported tick encounters in Switzerland for the period 2015–2020, collected in a citizen science project [8]. Each sample consists of a spatial estimation of the encounter by the volunteer, expressed as a circle with given spatial coordinates and radius that the volunteer thinks the encounter was within. In order to account for this uncertainty in the estimation of the spatial distribution of tick encounters, a Monte Carlo sampling procedure is applied. For each observation, we generate new samples by drawing from a 2d normal distribution centered at the circle center and using the radius as standard deviation. Furthermore, we create a spatial grid of 100×100 meters and count the observations in each cell. This data provides a sample of the *tick risk* distribution R, i.e., it represents the counts or the rate of human-tick encounters in each cell. R and H are connected via the *human exposure* E, i.e., the level of human activity with a susceptibility to tick encounters, which in turn depends on input factors e.g. population density and attractiveness [7]. In accordance with the literature [9], we assume that in any given cell we can express risk as a combination of exposure and hazard:

$$R = H \times E \tag{1}$$

The area of interest is all of Switzerland excluding water bodies, yielding 3.75 million cells.

2.2   The Predictive Model

Since observation data are only available for R, but we are primarily interested in H (and to some extent in E), we choose a composite model architecture akin to the one in Fig. 1. That is, we first define E and H as functions of their respective inputs X_E and X_H. For these inputs, we can use the inputs from our real-world data, or we can generate arbitrary features maps. The neural networks NN1 and NN2 can be seen as universal approximators for $E = f_1(X_E)$, $H = f_2(X_H)$. Each of these networks consists of an input layer followed by at least one hidden layer and an output layer. Because rates cannot be negative, we restrict these intermediate outputs to positive numbers by using the softplus activation function $f(x) = \ln(e^x + 1)$ in the intermediate output layers. The product of H_M and E_M is then used to compute the output R_M. The final output can be interpreted as the human-tick interaction rate and we can therefore model event counts by a Poisson distribution with $C \sim P(\lambda = R_M)$. Using real observation data, we can compute the likelihood of the data and optimize the model using a gradient descent algorithm. In practice, we train the model using the adam optimizer [10] and the negative log likelihood as loss function.

2.3 The Tick World Oracle

The oracle is a model to simulate data for H, E, R in pseudo-realistic scenarios. Furthermore, given the risk, we generate observation data by modelling the data collection process from the app users, so as to capture the uncertain nature in the data. The oracle outputs E_O and H_O are determined by two models with grid cell specific inputs (influence factors), X_E and X_H respectively, plus some noise terms ϵ_E, ϵ_H. The models are generally defined as functions g_1, g_2 of their parameter sets θ_H, θ_E and the influence factors:

$$E_O = g_1(X_E, \theta_E) + \epsilon_E, \ H_O = g_2(X_H, \theta_H) + \epsilon_H \tag{2}$$

In the simplest case, the literature (e.g., [5,6,11]) proposes regression models with the structure of a shallow neural network,

$$E_O = s(\theta_E \cdot X_E + \epsilon_E), \ H_O = s(\theta_H \cdot X_H + \epsilon_H), \tag{3}$$

where s is a suitable activation function restricting the rates to positive numbers, such as softplus. Naturally, we can increase the complexity of the oracle by using neural networks with hidden layers.

The risk R_O is calculated according to the logic above, i.e.,

$$R_O = E_O \times H_O \tag{4}$$

Again, the counts for human-tick encounters in a grid cell are simulated by drawing from the Poisson distributions $C \sim P(\lambda = R_O)$ in each cell. By scaling the rate parameters in each cell equally, we can generate an arbitrary number of simulated tick encounters, resulting in simulated count data for each cell. Each event is placed in its grid cell randomly, transforming the events to points with continuous space coordinates. The real tick observation data come with spatial reporting uncertainties. The app user choose a circular area on the map by zooming and moving to a specific location. The app then reports the center of the circle as the spatial coordinates and the its radius as the reported accuracy. The oracle simulates the uncertainty regarding the site of encounter by modelling the distance travelled between picking up a tick and noticing it. First, a site in distance d from the true location in a random direction is chosen. The distance d is drawn from an exponential distribution with a predefined scale μ_d (interpreted as the mean traveling distance between pickup and notice). The actually reported distance is then computed by scaling the real distance by a factor following a log normal distribution with median 1 and a predefined standard deviation σ_r, which can be interpreted as report quality of the 'citizen scientists'. New oracle scenarios ('oracle worlds') can be generated by varying the parameters $\theta_E, \theta_H, \epsilon_E, \epsilon_H, \mu_d, \sigma_r$.

2.4   Model Evaluation and Optimization

For evaluating the performance of a model M with outputs E_M, H_M, R_M in a given oracle scenario $O(\theta_E, \theta_H, \epsilon_E, \epsilon_H, \mu_d, \sigma_r)$, we compare the model outputs to the ground truths (E_O, H_O, R_O) of the zero noise version of the oracle scenario, that is the oracle scenario with $\epsilon_E = \epsilon_H = 0$. We use the Spearman rank correlations, yielding r_H, r_E, r_R.

While the correlations can be studied for each output individually, we also study the geometric mean $\bar{r}$ of them as a measure of overall performance. Furthermore, we want to investigate how robust our model is when dealing with noisy or feature-incomplete worlds and human sampling errors. For this purpose, we propose to average over noisy worlds, by varying the noise levels ϵ or their standard deviations σ respectively. Assuming Gaussian noise with $\epsilon(\mu = 0, \sigma)$, we may thus average over noise levels and write formally:

$$\bar{r}_\epsilon = \frac{1}{\sigma_E^{\max} \cdot \sigma_H^{\max}} \int_0^{\sigma_E^{\max}} \int_0^{\sigma_H^{\max}} \bar{r}(\epsilon_E, \epsilon_H) \; d\sigma_E \, d\sigma_H \tag{5}$$

The quantity $\bar{r}_\epsilon$ serves as a measure of fitness for a model and can be used to optimize the hyperparameters using an appropriate optimization scheme. Similar computations can be done by alternative varying the sampling simulation parameters μ_d and σ_r.

3   Preliminary Results

3.1   Oracle Worlds

For our preliminary studies we investigated oracle worlds with a shallow structure, as described in Eq. 3. We generated 5 random sets of different weight parameters θ_E, θ_H and varied the noise terms. For simplicity, we used Gaussian noise with $\mu = 0$ and the same variance for both noise terms in each run $\sigma_{\epsilon_E} = \sigma_{\epsilon_H}$, with $\sigma \in \{0, 1, 2, 3, 4, 5\}$. For the mean reported distance we chose $\mu_d = 500\,\mathrm{m}$ for all scenarios, in accordance with our real world data. The variance in distance estimation was set to $\sigma_r = 0.1$. This resulted in 30 distinct world scenarios (5 sets of weight parameters times 6 noise levels). We present visualizations of two worlds, both with the same weight parameters and different noise levels in Fig. 2. The first world has a zero noise term, representing an ideal case, in which our input features contain all the influence factors for exposure and hazard. The second world shown is one with the largest noise term ($\sigma = 5$).

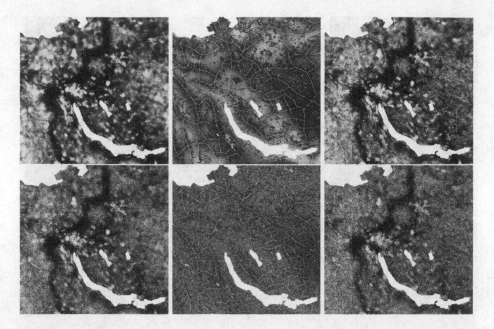

Fig. 2. Clean (top) and noisy (bottom) versions of Oracle worlds with the same weight sets for the Lake Zurich area. From Left to right: Exposoure, Hazard, Risk. Quantile Colorscale from low to high: Black-Magenta-Orange-Yellow (Color figure online)

3.2 Model Predictions

Predicted maps for both the clean and noisy worlds are visualized in Fig. 3. Upon visual inspection, both prediction sets resemble the clean world ground truth. Furthermore, both captured the broader coarse landscape to some degree, while struggling to recover the fine details visible in the hazard part of the oracle visualizations (c.f. Fig. 2). There are visible differences between the ground truth hazard and both model hazard outputs. Furthermore, the outputs from the noisy and clean model seem to differ. However, these differences are not visible when comparing both models risk outputs. The noisy version seems to show some artefacts (details not present in the oracle ground truth). The correlation coefficients are $r_E = 0.89, r_H = 0.74, r_R = 0.76$ for the clean version and $r_E = 0.80, r_H = 0.68, r_R = 0.76$ for the noisy version, supporting our visual observations.

3.3 Robustness

The geometric means for each oracle scenario are summarized in Table 1. As expected, our model's performance as measured by this metric seems to deteriorate with increasing noise levels, indicated by a decrease in prediction quality with increasing noise for each world. Furthermore, some world scenarios yield better performance than others.

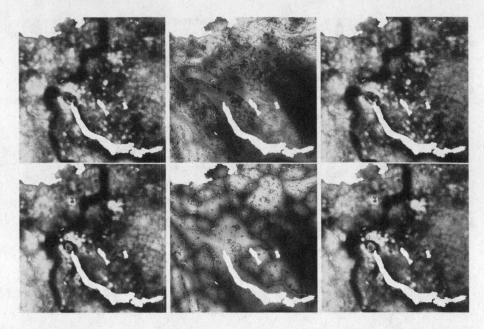

Fig. 3. Predictions from the clean (top) and noisy scenarios (bottom) with otherwise identical parameters. From Left to right: Exposoure, Hazard, Risk. Colorscale from low to high: Black-Magenta-Orange-Yellow (Color figure online)

4   Discussion and Outlook

Our preliminary results show how to generate oracle world scenarios for the validation of an underconstrained model with no known ground truth. Within our approach, our preliminary results suggest that our modelling pipeline can produce useful predictions, even in noisy scenarios. Specifically, we were able to reproduce the coarse structure of oracle worlds using our predictive models. Nonetheless, the predictive model failed to recover finer details, which could

Table 1. Geometric mean of Spearman correlation coefficients between model outputs and ground truth for different world scenarios.

Parameter Set (θ_E, θ_H)	Noise (σ)					
	0	1	2	3	4	5
a	.79	.79	.77	.76	.74	.77
b	.81	.80	.80	.78	.79	.79
c	.83	.85	.82	.84	.82	.82
d	.78	.75	.77	.76	.75	.73
e	.79	.79	.80	.80	.79	.77

represent an upper bound in terms of achievable performance of the model and might have important implications for our real world application, for example when choosing the detail level of our final predictions. Future work should investigate what spatial properties are required for our inputs such that our model is able to learn their contribution from the uncertain and noisy data. This could help in further feature selection and engineering.

Our preliminary results represent a first proof of principle. The full procedure, as outlined above, is still work in progress. In a next step, it remains to be demonstrated that our approach is also able to reproduce more complex oracle worlds using our hyperparameter optimization approach. In a final step, the oracle itself must be systematically compared with real world data to establish a solid level of confidence for our predictions. After establishing that our approach selects a model of adequate complexity for our toy worlds, the next step will be to estimate how close our toy worlds are to the real world, or in other words we will have to find a measure of confidence in the oracle itself.

References

1. Rykiel, E.J.: Testing ecological models, the meaning of validation. Ecol. Model. **90**(3), 229–244 (1996). https://doi.org/10.1016/0304-3800(95)00152-2
2. Barr, E.T., Harman, M., McMinn, P., Shahbaz, M., Yoo, S.: The Oracle problem in software testing: a survey. IEEE Trans. Softw. Eng. **41**(5) (2015). https://doi.org/10.1109/TSE.2014.2372785
3. Prastawa, M., Bullitt, E., Gerig, G.: Synthetic ground truth for validation of brain tumor MRI segmentation. In: Duncan, J.S., Gerig, G. (eds.) MICCAI 2005. LNCS, vol. 3749, pp. 26–33. Springer, Heidelberg (2005). https://doi.org/10.1007/11566465_4
4. Bundesamt für Statistik (BFS): Arealstatistik Standard nach Nomenklatur 2004 https://www.bfs.admin.ch/hub/api/dam/assets/4103539/master. Accessed 13 July 2021
5. Boehnke, D., et al.: Estimating Ixodes ricinus densities on the landscape scale. Int. J. Health Geograph. **14**(23) (2015). https://doi.org/10.1186/s12942-015-0015-7
6. Willibald, F., van Strien, M.J., Blanco, V., Grêt-Regamey, A: Predicting outdoor recreation demand on a national scale-The case of Switzerland. Appl. Geogr. **113** (2019). https://doi.org/10.1016/j.apgeog.2019.102111
7. Zeimes, C.B., Olsson, G.E., Hjertqvist, M., Vanwambeke, S.O.: Shaping zoonosis risk: landscape ecology vs. landscape attractiveness for people, the case of tick-borne encephalitis in Sweden. Parasites Vectors **7**(1), 1–10 (2014). https://doi.org/10.1186/1756-3305-7-370
8. App 'Tick Prevention'. https://zecke-tique-tick.ch. Accessed 13 July 2021
9. Garcia-Marti, I., Zurita-Milla, R., Swart, A.: Modelling tick bite risk by combining random forests and count data regression models. PLOS ONE **14**(12) (2019). https://doi.org/10.1371/journal.pone.0216511
10. Kingma, D., Ba, J.: Adam: a method for stochastic optimization. arXiv:1412.6980
11. Del Fabbro, S., Gollino, S., Zuliani, M., Nazzi, F.: Investigating the relationship between environmental factors and tick abundance in a small, highly heterogenous region. J. Vector Ecol. **40**(1), 107–116 (2015). https://doi.org/10.1111/jvec.12138

Droplets, Fluids, and Synthetic Biology

Obstacles on the Pathway Towards Chemical Programmability Using Agglomerations of Droplets

Johannes Josef Schneider[1]([✉])(iD), Alessia Faggian[2](iD), Hans-Georg Matuttis[3], David Anthony Barrow[4](iD), Jin Li[4](iD), Silvia Holler[2](iD), Federica Casiraghi[2], Lorena Cebolla Sanahuja[2], Martin Michael Hanczyc[2,5](iD), Patrik Eschle[1](iD), Mathias Sebastian Weyland[1](iD), Dandolo Flumini[1](iD), Peter Eggenberger Hotz[1], and Rudolf Marcel Füchslin[1,6](iD)

[1] Institute of Applied Mathematics and Physics, School of Engineering, Zurich University of Applied Sciences, Technikumstr. 9, 8401 Winterthur, Switzerland
`johannesjosefschneider@googlemail.com`,
`{scnj,escl,weyl,flum,eggg,furu}@zhaw.ch`
[2] Laboratory for Artificial Biology, Department of Cellular, Computational and Integrative Biology, University of Trento, Via Sommarive, 9, 38123 Povo, Italy
`{alessia.faggian,silvia.holler,federica.casiraghi,lorena.cebolla,`
`martin.hanczyc}@unitn.it`
[3] Department of Mechanical Engineering and Intelligent Systems, The University of Electrocommunications, Chofu Chofugaoka 1-5-1, Tokyo 182-8585, Japan
`hg@mce.uec.ac.jp`
[4] Laboratory for Microfluidics and Soft Matter Microengineering, Cardiff School of Engineering, Cardiff University, Queen's Buildings, 14-17 The Parade, Cardiff CF24 3AA, Wales, UK
`{Barrow,LiJ40}@cardiff.ac.uk`
[5] Chemical and Biological Engineering, University of New Mexico, MSC01 1120, Albuquerque, NM 87131-0001, USA
[6] European Centre for Living Technology, Ca' Bottacin, Dorsoduro 3911, Calle Crosera, 30123 Venice, Italy
`https://www.zhaw.ch/en/about-us/person/scnj/`

Abstract. We aim at planning and creating specific agglomerations of droplets to study synergic communication using these as programmable units. In this paper, we give an overview of preliminary obstacles for the various research issues, namely of how to create droplets, how to set up droplet agglomerations using DNA technology, how to prepare them for confocal microscopy, how to make a computer see droplets on photos, how to analyze networks of droplets, how to perform simulations mimicking experiments, and how to plan specific agglomerations of droplets.

Keywords: Chemical programmability · Artificial life · Droplets · Agglomeration · Computer vision · Network analysis · Simulation

This work has been partially financially supported by the European Horizon 2020 project *ACDC – Artificial Cells with Distributed Cores to Decipher Protein Function* under project number 824060.

The original version of this chapter was previously published non-open access. A Correction to this chapter is available at https://doi.org/10.1007/978-3-031-23929-8_25

1   Introduction

Usually, when writing a paper one starts off with describing the problem one considers, how it is embedded in a whole class of problems, why it is of importance, and cites own preliminary work and the related papers of others in this field. Then one describes the problem in a more detailed way and also the approach one intends to use on this problem. Afterwards one presents results achieved with this approach and rounds the paper off with a summary of the results and an outlook on future work. In this way, this paper is highly unusual, as we present a list of obstacles on the pathway to what we intend to do. This list is most probably even not complete, these are only the problems we have become aware of in recent years. We have made progress with all of them but are still far from solving many of them in the way we intend.

Within the context of the EU Horizon project ACDC (*Artificial Cells with Distributed Cores to Decipher Protein Function*), we aim at developing a chemical compiler [4,27]. This device should be able to e.g. produce some desired macromolecules on demand, especially for medical purposes. It contains a large variety of chemical educts, which are filled in droplets of various sizes. The task is to create a specific agglomeration of some of these droplets. They should then exchange chemicals in a controlled way via pores in the bilayers formed during the agglomeration process, thus allowing a gradual chemical reaction scheme, which then efficiently produces the desired macromolecule.

In order to get to the point where we can design an experiment leading to a desired agglomeration of droplets, we need to know

- how to best create droplets of specific sizes,
- how this agglomeration process works
  - by setting up various experiments leading to an agglomeration of droplets,
  - by taking pictures during and after the agglomeration process,
  - by analyzing these pictures on a computer,
  - by analyzing the networks of droplets within the agglomerations,
  - and then by iteratively changing experimental parameters and analyzing the resulting changes in the networks,
- what the physical laws governing the agglomeration process are
  - by modelling the agglomeration process [20],
  - by simulating this model on a computer [22], mimicking an experiment,
  - by analyzing the networks of droplets resulting in these simulations [20, 21],
  - by comparing these results with the corresponding outcomes from the experiments,
  - and then by iteratively changing the model in order to get closer to the experimental outcome,
- what specific agglomeration of droplets is needed for producing the desired macromolecule
  - by planning the gradual reaction scheme resulting in the desired macromolecule,

- by planning the reaction network allowing this gradual reaction scheme,
- by being aware of geometric and physical limitations to realize specific networks,
- by embedding the network of droplets for the reaction process in an even larger agglomeration of droplets to physically stabilize the central reaction network if necessary,
- and by designing the experiment leading to this specific agglomeration of droplets.

2   Setting Up the Experiment

2.1   Creating Droplets

Different ways of creating droplets have their specific advantages and drawbacks. In the experiments performed in David Barrow's group at Cardiff University, T-junctions are used in which a fluid stream inside another fluid is split in a sequence of droplets. With this microfluidic approach, it is necessary to determine the correct pressure ratio between the two fluids, such that the system is in a dripping and not a jetting or squeezing regime [10]. In the T-junction, the continuous phase (watery) fluid breaks the dispersed (oily) phase fluid, and consequently the dispersed phase fluid is sheared to form tiny droplets. This geometry is easy to fabricate and creates droplets of relatively high monodispersity at a high rate.

Another approach is used in experiments performed in Martin Hanczyc's group at the University of Trento. Hereby a low-concentrated solution of phospholipids in oil, which is first sonicated for 15 min, is added to an aqueous hosting solution with low concentrations of glucose, sodium jodide, and Hepes. Some dodecane is added to avoid unwanted chloroform production and evaporation. As the density of the heavy oil used is larger than the density of water, the oil sinks through the water to the bottom of the phial. Then this mixture is emulsified by mechanical agitation, i.e., by shaking and repeatedly rubbing the phial over a vibrating board. While the droplets created in Cardiff exhibit almost the same radius value, this approach used in Trento leads to a polydisperse system of emulsion droplets surrounded by hulls of phospholipids with lognormally distributed radii [6].

2.2   Surface Functionalization and Self-assembly Process

While the droplets in the experiments in Cardiff can immediately form more or less extended bilayers when touching each other, thus automatically starting an assembly process, the lipid surfaces of the droplets in Trento must be functionalized to allow connections with other droplets. For this purpose, streptavidin attached to a red or green fluorescent label and ssDNA oligonucleotides are diluted in the hosting solution and preincubated in the dark for 30 min at room temperature. (The sequences and the modifications of the ssDNA oligonucleotides are shown in [5].) Then this solution is mixed with the solution containing the droplets and incubated in the dark on a rocking at 40-60 rpm for roughly one hour, resulting in different droplet populations depending on streptavidin-DNA specific composition. Then the droplets are washed twice by centrifugation, by removing supernatant with vacuum, and replacing it with a fresh hosting solution. These functionalized droplets are then used immediately for the experiment. As shown in Fig. 1, the type of connections which can be formed by the droplets can be restricted and controlled by using complementary strands of DNA. Two droplet populations with different streptavidin colors are together incubated for two or three hours at room temperature in the dark and transferred afterward to a sealed microscopy observation chamber by gentle aspiration without resuspension.

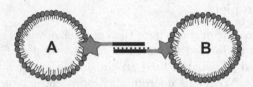

Fig. 1. Sketch of a pair of oil-filled droplets in water, to which complementary strands of ssDNA oligonucleotides are attached: The surfaces of the droplets are composed by single-tail surfactant molecules like lipids with a hydrophilic head on the outside and a hydrophobic tail on the inside, thus forming a boundary for the oil-in-water droplet. By adding some single-strand DNA to the surface of a droplet, it can be ensured that only desired connections to specific other droplets with just the complementary single-strand DNA can be formed. Please note that the connection of the droplets in this picture is overenlarged in relation to the size of the droplets. In reality, the droplets have a radius of 1–50 µm, whereas a base pair of a nucleic acid is around 0.34 nm in length [1], such that the sticks of connecting DNA strands are roughly 5 nm long.

2.3   Taking Pictures

Before pictures can be taken, the microscope slides are prewashed sequentially with acetone and ethanol and are dried with compressed air. Then they are covered with imaging spacers. The microscope chamber is preloaded with a mixture of the aqueous solution and some sample of the agglomeration. Finally, the observation chamber is effectively sealed with a prewashed coverslip. The samples at the end-point are evaluated using spinning disk confocal microscopy, as shown in Fig. 2.

3   How to Make a Computer "See" Droplets

When getting a stack of pictures of an agglomeration of red and green droplets, one needs to derive the information about the locations of their midpoints,

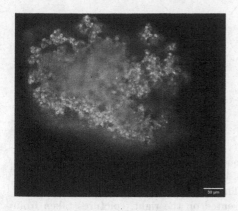 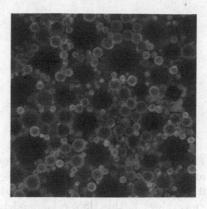

Fig. 2. Left: Picture of an assembly taken with confocal microscopy. Additionally to the red and green "painted" droplets, a yellow dot lights up at each DNA connection between pairs of red and green droplets. Right: Small part of an agglomeration of a polydisperse system of red and green droplets. (Please note that some of the circum-circles of larger droplets are displayed diffuse.)

their radii, and the connections between them. Each picture contains a two-dimensional cut through the three-dimensional agglomeration, as shown in Fig. 2. While a human can easily discern these droplets, several steps are needed to derive the information about positions and radii of the droplets on a computer, on which each picture is usually comprised by a two-dimensional array of pixels.

3.1   Color Filtering

Each pixel contains a color information. Usually, the RGB color model is used, i.e., for each pixel, there is one value for red, one value for green, and one value for blue. For example, if one byte is used for each of these three colors, then we have values between 0 and 255 for each of them, and thus more than 16.7 million colors can be represented on a computer.

However, the problem is that most colors are mixed colors with large proportions of at least two color channels. Thus, when using the RGB color model for filtering colors on a picture like the one in Fig. 2, while hoping to get either red or green, in many cases one will get signals for both colors and one then has to check whether there is a more dominant value in the other channel. Thus, it is preferrable to work with another color model, in which the color is defined by one value only. An example for such a model is the HSV color model, in which the possible colors are also represented by three values. But the true color information is stored in the so-called hue color-angle H only, as shown in Fig. 3. The other two values are the saturation S and the darkness value V. Each color stored in a RGB color coding can also be represented in this HSV color model (and related models) and vice versa [29]. But in this model, the filtering process

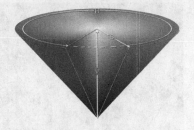
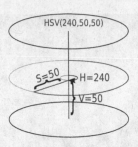

Fig. 3. HSV color model: On the left, the cone of possible colors for various hue color angles H, saturations S, and darkness values V are shown. H can take values between $0°$ and $360°$, S and V between 0 and 100. The example of how to compose the color "navyblue" in this HSV color model is presented on the right. (pictures taken from wikipedia [29]) (Color figure online)

is much easier, as one has to apply a filter for a specific range of hue angles only, only keeping in mind that S and V should have appropriate values. After this filtering for either red or green, one has two black and white pictures, in which the white and grey pixels correspond to pixels of formerly one color angle range, while the pixels in other color angle ranges are represented in black now.

3.2 Detecting Edges

In the next step, one has to determine the boundaries of a droplet. As can be seen in Fig. 2, some droplets appear as boundary rings only, others are completely filled disks, but most droplets are shown with a more or less sharp ring and an interior with a smaller saturation and a larger darkness value of the same color. In order to determine the boundaries of a droplet, an edge detection algorithm has to be applied.

A simple example for edge detection (Please note that the word "edge" in this section describes a part of a boundary of an object, whereas we use the word edge as a connection between a pair of nodes when referring to networks otherwise in this paper.) is the Sobel operator [23,30]. Let A be a matrix containing the pixels of a color filtered picture, i.e., $A(i,j)$ only contains a value for the greytone of this pixel, e.g., a value between 0 and 255. Then one applies

$$G_x = \begin{pmatrix} 1 & 0 & -1 \\ 2 & 0 & -2 \\ 1 & 0 & -1 \end{pmatrix} * A \text{ and } G_y = \begin{pmatrix} 1 & 2 & 1 \\ 0 & 0 & 0 \\ -1 & -2 & -1 \end{pmatrix} * A. \quad (1)$$

Thus, for a specific pixel (x,y), $G_x(x,y)$ stores the value $A(x-1,y-1) - A(x+1,y-1) + 2A(x-1,y) - 2A(x+1,y) + A(x-1,y+1) - A(x+1,y+1)$. The larger the absolute value of $G_x(x,y)$, the more pronounced in the transition from black to white or vice versa in x-direction. The G_y operator analogously detects such transitions in the y-direction. They can be combined to $G = \sqrt{G_x^2 + G_y^2}$, thus considering transitions in both x- and y-direction.

Here the next difficulty occurs: one has to find a proper cutoff, i.e., an answer to the question which $G(x, y)$ value is large enough to show the transition from the outside of a droplet to its boundary. Thus, one defines a matrix B with $B(x, y) = 1$ if $G(x, y)$ exceeds some threshold value and $B(x, y) = 0$ otherwise. However, larger droplets are often bounded by a rather diffuse ring only, such that the corresponding G values are not large. Thus, when choosing a too large threshold value, one will miss out many droplets, whereas a too small value will lead to many concentric rings and also to ghost particles which are not there. More elaborate edge detection operators have been developed in recent years [15], like the weights $(47, 162, 47)$ instead of $(1, 2, 1)$ [30] or weighted sums over even more pixels, but one still faces the problem to manually determine appropriate threshold values.

3.3   Identifying Objects

So far, the computer only has a matrix B of true/false values whether a transition takes place at a specific pixel or not, but it does not know anything about droplets and enclosing rings yet. Using the Hough transform [3,8], structures like lines or circles can be detected. This method can be most easily explained for the detection of lines, as shown in Fig. 4.

For other structures like circles, the Hough transform can also be applied in an altered way. In contrast to a line which can be described by a slope and a y-axis-section only, three parameters are needed for a circle, namely the coordinates (x_m, y_m) for the midpoint and the radius r, such that a histogram over three dimensions has to be created. In order to focus the search for correct radii, the range of radius values is restricted a priori. Then one performs a loop over possible x_m values with some stepwidth and within this loop, a further loop over possible radii r with some stepwidth, calculates in the innermost third loop over all pixels the possible corresponding y_m values if $|x_{\text{pixel}} - x_m| \leq r$ and increments the corresponding entries in the three-dimensional histogram. Alternatively, one can go through all triples of pixels, assume that they might lie on a circle, and derive its midpoint coordinates and radius from them, as three points are necessary to define a circle. The peaks in the histogram indicate the circles and contain the information about their midpoints and radii. However, like for the edge detection, one has also to find appropriate threshold values for detecting true circles, while keeping in mind that the peaks for the small circles need to be lower than those of larger circles. Due to the diffuse nature of many larger circles and also because of other problems, especially the huge number of ten-thousands of droplets in one picture, an approach trying to detect all droplets and to determine their midpoint coordinates and their radii simultaneously only with the described methods fails. Instead, a very elaborate algorithm to detect droplets in a successive way has to be developed with a commercial software [11].

So far, we have only considered the evaluation of one picture only. But as we are interested in three-dimensional structures, a whole stack of pictures is needed.

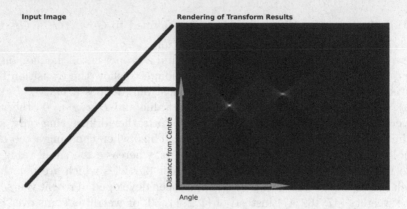

Fig. 4. Detection of lines in a picture using the Hough transform: On the left side, an exemplary picture with two lines is shown. A line can be represented by its Hesse normal form as $r = x_{\text{pixel}} \cos \vartheta + y_{\text{pixel}} \sin \vartheta$, with r being the shortest distance from a point on the line to the origin and ϑ being the angle between the line connecting this closest point to the origin and the x-axis. If scanning the range of possible ϑ values with small steps, performing a loop over all pixels for each value of ϑ, calculating the corresponding r value, and creating a histogram of (ϑ, r) values as shown on the right, one finds two peaks in this example, one peak for each line. Thus, this method returns the information about these two lines. (picture taken from wikipedia [28])

At least two cuts through each droplet are necessary in order to determine its real radius value and a third one to exclude the error that data from two different droplets have been mixed up. Here further problems occur, as the droplets move between the taking of successive pictures in a picture stack and as the microscope automatically readjusts the brightness when taking a new picture.

4   How to Analyze Networks of Droplets

While these algorithms for edge detection and object identification work very well on drawings or pictures of technical devices, they are often unable to detect the boundaries of a face on a painting, and also for our case, we have severe difficulties as already mentioned. But let us now assume that we have finally mastered the problem of detecting all droplets properly. As a next step, we have to determine the network of droplets. Mathematically speaking, a network (sometimes called graph) is comprised of a set of nodes and a set of edges connecting pairs of nodes. A network can be represented by an adjacency matrix η between pairs (i, j) of nodes with

$$\eta(i, j) = \begin{cases} 1 & \text{if an edge between } i \text{ and } j \text{ exists} \\ 0 & \text{otherwise} \end{cases} \tag{2}$$

The adjacency matrix contains the whole information about the network.

In our case, the nodes are identical with the spherical droplets. An edge exists if the distance between the midpoints of two droplets is smaller than the sum of their radii, i.e.,

$$\sqrt{(x_i - x_j)^2 + (y_i - y_j)^2 + (z_i - z_j)^2} \leq r_i + r_j + \theta \tag{3}$$

with some threshold value θ, which needs not only to be introduced because of numerical inaccuracy but also in order to consider the short DNA strands connecting pairs of droplets, and if some further constraints are met, e.g., that only connections between red and green droplets can be formed [20]. Please note that θ can be negative and depend on r_i and r_j if the droplets lose their spherical shapes in the case of extended bilayer formation.

4.1   Analyzing one Network

One can take different viewpoints when analyzing such a network: On the local or elementary level, one considers the properties of the nodes, e.g., the degree $D(i)$ of a node i, which is the number of connections node i has to other nodes and which can be easily derived from the adjacency matrix by $D(i) = \sum_j \eta(i,j)$. On an intermediate level, groups of nodes are considered like the question what the maximum clique size is, i.e., the maximum number of nodes in a subset of the graph which are pairwise connected with each other. From a global viewpoint, the properties of the overall network are studied. One such property is the maximum of the geodesic distances between pairs of nodes. The geodesic distance $d(i,j)$ between two specific nodes i and j is the minimum number of edges which have to be trespassed to get from i to j. Another widely studied question is whether the network is contiguous or whether it is split in a number of clusters, how large the number of these clusters is, and how large the size of the largest cluster is [24]. Please note that we refer to a cluster as a subset of nodes in which each node can be reached from any other node in the same cluster by using only edges connecting the nodes of this cluster. In contrast, a clique is defined as a subset of nodes in which all of these nodes are directly connected with each other by an edge. Finally, one can also take a local-global viewpoint and e.g. study the importance of specific nodes for the overall network [16]. Figure 6 shows exemplary results for the time evolution of some of these observables during a simulation.

4.2   Comparing Networks

So far, we have only analyzed networks separately. Of course, we can then perform statistics of the properties of these networks and can have a look at how these statistics change if experimental parameters are altered. However, for our purposes, we want to generate exactly the same networks or at least networks with equal parts again and again. Thus, we have also to compare networks. But already finding out whether two networks are identical is far from being a trivial task as shown in Fig. 5.

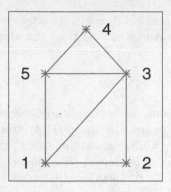 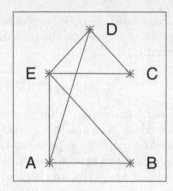

Fig. 5. Toy example for the graph isomorphism problem: The two networks shown above are comprised of 5 nodes and 7 edges each and are seemingly different. But one can map them onto each other by the isomorphism $E = 3, A = 5, B = 4, D = 1$, and $C = 2$.

Usually, no isomorphisms between entire networks can be found, such that one has to restrict oneself to the detection of common structures like a common subnetwork. Then one can group the various networks according to the commonly detected subnetworks. For this purpose, we intend to e.g. extend the k-shell decomposition algorithm [2] for determining whether a common network nucleus exists and which fractal subcomponents exist in various configurations.

5   How to Understand, Simulate, and Plan the Agglomeration Process

5.1   Simulating the Agglomeration

So far, we have only spoken about the experiments performed and the analysis of the experimental outcomes. But in order to truly understand the agglomeration of droplets, we have to develop a simplified model, simulate this model on a computer, thus mimicking the experiment, and compare the results of the simulations with the experimental findings. If they deviate, then the model has to be adjusted or perhaps even another model has to be developed. However, each model has to be built on physical laws. So, each droplet sinks in water due to gravity reduced by the buoyant force. Due to its spherical shape, we also apply the Stokes friction force $\boldsymbol{F}_{i,\mathrm{S}} = -6\pi\eta r_i \boldsymbol{v}_i$ to it as well as the added mass concept [25]. However, these last laws do not consider the hydrophilic nature of the surface of the droplet, which might slightly alter these laws. Furthermore, we add some small random velocity changes to simulate the motion induced in the experimental setup, which cannot be modeled exactly, and almost-elastic collision laws if droplets touch each other or the surface.

Figure 6 presents the time evolution of some observables in network analysis. For the simulations, the same parameters were used as in [20] with 2,000 droplets initialized in an inner part of a cylinder only ("narrow initialization"). This leads to significant qualitative and quantitative differences to results presented in [19], in which a "wide initialization" routine was implemented, for which the whole cylinder volume was used.

Of course, these simulation results will not coincide exactly with experimental findings. However, in an iterative process with analyzing pictures and movies from experiments, the model parameters have to be gradually changed. Then a simulation finally mimicks an experiment much better than in our current results shown in Fig. 7, perhaps not in the exact trajectories of the various particles, but in the statistics created of all particles.

5.2   Planning the Gradual Reaction Scheme

The theoretical and analytical work so far followed a bottom-up approach by starting either from physical laws for modelling and simulating the experimental setup or starting from pictures taken from experimental outcomes which are then analyzed. On the other hand, we would also like to follow a top-down approach by starting at the desired results and then gradually taking steps down in order to better plan these last steps towards this goal, which later on shall become the first steps.

As already mentioned, among other aims, we intend to produce macro-molecules using the technology of a chemical compiler we intend to develop. For this purpose, we first need to know how to produce these macromolecules in order to plan the gradual reaction scheme. For many pharmaceutical macromolecules, various ways are known how to synthesize them [9]. Using small droplets, we think that further methods to produce these macromolecules could be developed as the compartimentalization of usually semi-stable molecules might stabilize them, thus offering new possibilities for synthesis methods.

5.3   Designing the Arrangement of Droplets with Geometric and Physical Restrictions

The plan of the gradual reaction scheme has to be translated into a network of droplets. Trivially, the nodes represent the droplets being filled with the educts. But already at this point, one has to consider which relative quantities of educts and thus which relative radii of droplets are needed. Then there are edges between those nodes between which there are chemical reactions during the first step of the gradual reaction scheme. But then the question arises how to move on. There are various ways conceivable of how to implement the second step of the gradual reaction scheme. For example, those groups of droplets which together produced some intermediary product in the first step could be

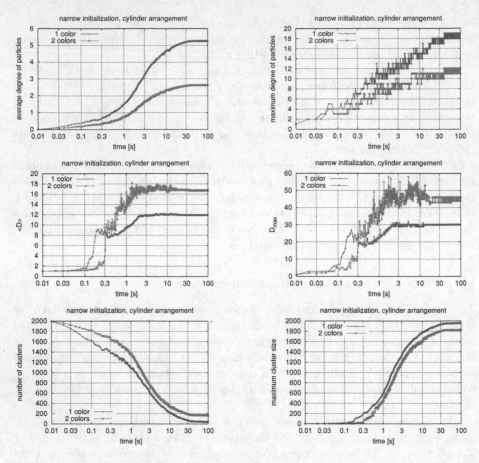

Fig. 6. Time evolution of network observables during a simulation (see Sect. 5) of a polydisperse system of droplets with two types of particles (curves marked as "2 colors": "red" R and "green" G): only $R - G$ connections can be formed but not $R - R$ and not $G - G$. For comparison, also the results for the scenario that only one type of particles exists are shown (curves marked as "1 color"). Top: mean value (left) and maximum (right) of degrees of the droplets. Middle: mean value (left) and maximum (right) of geodesic distances between pairs of droplets. Bottom: number of clusters (left) and maximum cluster size (right).

enclosed in a further hull into which they emit the resulting molecules of this first reaction step. Another possibility would be that these groups of droplets unite to one larger droplet. In these two cases, these groups of droplets are then represented by a super-node, and the super-nodes are then connected with edges representing chemical reactions between them in the second step of the gradual reaction scheme. This approach could then be iterated for the remaining steps. But a third possibility could be that the intermediary product of a chemical reaction could be found preferably in one of the droplets only. Then its repre-

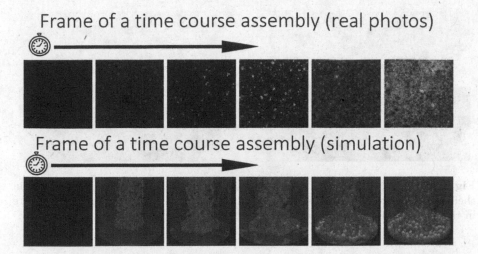

Fig. 7. Time evolution of an experimental agglomeration process (top row) and of a computer simulation trying to mimick the experiment (bottom row)

senting node has to be connected by edges with reaction partners in the next reaction step. Thus, in this approach, we will have edges which are typically only active for one reaction step, resulting in an adjacency matrix $\eta(i, j, t)$ with (i, j) denoting the pair of droplets between which a chemical reaction shall take place and t denoting the time step at which this chemical reaction shall be performed in the gradual reaction scheme. In the experiment, additional difficulties arise to activate or deactivate specific connections for specific reaction steps.

In order to translate the network of droplets into a real agglomeration, it might not only be necessary

- to work with droplets of different sizes, referring to the relative amount of chemicals needed for the reaction process, but also of different weights, i.e., to work with droplets filled with oils of differing densities, in which the chemicals are hosted,
- to work with various types of sticks on droplet surfaces, such that e.g. not only "$A - B$" connections but also "$A - C$" connections can be formed,
- to develop sticks beyond the concept of complementary ssDNA strands, so that connections for successive steps in the gradual reaction scheme can be formed.

However, for such a reaction network and a corresponding agglomeration of droplets, physical and geometric restrictions have to be taken into account. For some of these restrictions, one can exploit findings from related areas of research. For example, some decades ago Newton's solution to the so-called kissing number problem was proved to be correct, according to which only up to 12 spheres of equal radius can touch a further sphere of the same radius in their midst without overlaps [26]. Another related problem, the question what the maximum number

Fig. 8. While up to 12 spheres of equal radius can touch a sphere of the same radius in their midst without overlaps, already 28 spheres of equal radius can touch the sphere in their midst if its radius is twice as large [18].

of connections in a set of N sticky spheres of equal radius can be has been solved for small values of N [7]. However, we have to take into account that our droplets differ in size, such that we do not deal with a monodisperse but a polydisperse system. Already for other problems like the "circles in a circle" problem [14], it has been shown that the polydispersity significantly changes the results [12] and the nature [17] of this problem. In order to get better estimates for extreme values of a polydisperse system of touching spheres, we developed a heuristic for a bidisperse kissing number problem [18], which e.g. led to the conclusion that up to 28 spheres of radius 1 can touch a sphere of radius 2 in their midst, as shown in Fig. 8, thus providing a maximum number of kissing spheres in a polydisperse system in which the maximum radius is only twice as large as the smallest one.

And, finally, we have to find out how to design the experiment in a way that it indeed produces the desired agglomeration of droplets with the reaction network within, which in turn runs the gradual reaction scheme leading to the desired molecules. For this final step, we have to combine all findings from all the steps described above.

6   Conclusion and Outlook

In this paper, we have described several obstacles we have met on our pathway to build a chemical compiler. It shall be able to foretell how an experiment has to be designed in order to generate an agglomeration of droplets with which a gradual chemical reaction scheme can be enabled for the purpose of producing some specific macromolecule. But it can also be helpful to answer basic questions for the origin of life [13], e.g., which molecules could be produced with which probability if their educts were not simply mixed in a primordial soup but compartementalized in droplets which are then brought in contact with each other. This technology could also find an application as environment or medical biosensor. A further application might be the investigation and reproduction of metabolic catalysis in living beings as well as the creation of their counterparts for artificial lifeforms.

But before thinking of possible applications of such a device, we first have to find solutions for various problems. We have to find a reliable and reproducable way of generating droplets of various specific sizes, have to develop a method to automatically analyze pictures taken from their agglomeration, have to change our models for this agglomeration process, such that they mimick the experiments, and have to check with tools from network analysis whether we indeed generate the desired gradual reaction network as intended.

Acknowledgment. A.F. would like to thank Giorgina Scarduelli and the other members of the imaging facility group at the University of Trento.

References

1. Annunziato, A.T.: DNA packaging: nucleosomes and chromatin. Nat. Educ. **1**, 26 (2008)
2. Carmi, S., Havlin, S., Kirkpatrick, S., Shavitt, Y., Shir, E.: A model of internet topology using k-shell decomposition. Proc. Natl. Acad. Sci. **104**(27), 11150–11154 (2007)
3. Duda, R.O., Hart, P.E.: Use of the Hough transformation to detect lines and curves in pictures. Commun. ACM **15**, 11–15 (1972)
4. Flumini, D., Weyland, M.S., Schneider, J.J., Fellermann, H., Füchslin, R.M.: Towards programmable chemistries. In: Cicirelli, F., Guerrieri, A., Pizzuti, C., Socievole, A., Spezzano, G., Vinci, A. (eds.) WIVACE 2019. CCIS, vol. 1200, pp. 145–157. Springer, Cham (2020). https://doi.org/10.1007/978-3-030-45016-8_15
5. Hadorn, M., Boenzli, E., Hanczyc, M.M.: Specific and reversible DNA-directed self-assembly of modular vesicle-droplet hybrid materials. Langmuir **32**, 3561–6 (2016)
6. Hadorn, M., Boenzli, E., Sørensen, K.T., Fellermann, H., Eggenberger Hotz, P., Hanczyc, M.M.: Specific and reversible DNA-directed self-assembly of oil-in-water emulsion droplets. Proc. Natl. Acad. Sci. **109**(50), 20320–20325 (2012)
7. Hayes, B.: The science of sticky spheres - on the strange attraction of spheres that like to stick together. Am. Sci. **100**, 442 (2012)
8. Hough, P.V.: Method and means for recognizing complex patterns. U.S. Patent, pp. 3,069,654, 18 December 1962
9. Kleemann, A., Engel, J., Kutscher, B., Reichert, D.: Pharmaceutical Substances: Syntheses, Patents and Applications of the most relevant APIs, 5th edn. Thieme (2008)
10. Li, J., Barrow, D.A.: A new droplet-forming fluidic junction for the generation of highly compartmentalised capsules. Lab Chip **17**, 2873–2881 (2017)
11. Matuttis, H.G., et al.: The good, the bad, and the ugly: droplet recognition by a "shootout"-heuristics (2021). Submitted to the proceedings of the XV Wivace workshop, 15–17 September 2021, Winterthur (2021)
12. Müller, A., Schneider, J.J., Schömer, E.: Packing a multidisperse system of hard disks in a circular environment. Phys. Rev. E **79**, 021102 (2009)
13. Oparin, A.: The Origin of Life on the Earth, 3rd edn. Academic Press, New York (1957)

14. Pirl, U.: Der Mindestabstand von n in der Einheitskreisscheibe gelegenen Punkten. Math. Nachr. **40**, 111–124 (1969)
15. Scharr, H.: Optimal filters for extended optical flow. In: Jähne, B., Mester, R., Barth, E., Scharr, H. (eds.) IWCM 2004. LNCS, vol. 3417, pp. 14–29. Springer, Heidelberg (2007). https://doi.org/10.1007/978-3-540-69866-1_2
16. Schneider, J., Kirkpatrick, S.: Selfish versus unselfish optimization of network creation. J. Stat. Mech: Theory Exp. **8**, P08007 (2005)
17. Schneider, J.J., Müller, A., Schömer, E.: Ultrametricity property of energy landscapes of multidisperse packing problems. Phys. Rev. E **79**, 031122 (2009)
18. Schneider, J.J.: Geometric restrictions to the agglomeration of spherical particles. In: Schneider, J.J., et al. (eds.) WIVACE 2021, CCIS, vol. 1722, pp. 72–84. Springer, Cham (2022). https://doi.org/10.1007/978-3-031-23929-8_7
19. Schneider, J.J., et al.: Analysis of networks formed during the agglomeration of a polydisperse system of spherical particles (2021, submitted)
20. Schneider, J.J., et al.: Influence of the geometry on the agglomeration of a polydisperse binary system of spherical particles. In: ALIFE 2021: The 2021 Conference on Artificial Life (2021). https://doi.org/10.1162/isal_a_00392
21. Schneider, J.J., Weyland, M.S., Flumini, D., Füchslin, R.M.: Investigating three-dimensional arrangements of droplets. In: Cicirelli, F., Guerrieri, A., Pizzuti, C., Socievole, A., Spezzano, G., Vinci, A. (eds.) WIVACE 2019. CCIS, vol. 1200, pp. 171–184. Springer, Cham (2020). https://doi.org/10.1007/978-3-030-45016-8_17
22. Schneider, J.J., Weyland, M.S., Flumini, D., Matuttis, H.-G., Morgenstern, I., Füchslin, R.M.: Studying and simulating the three-dimensional arrangement of droplets. In: Cicirelli, F., Guerrieri, A., Pizzuti, C., Socievole, A., Spezzano, G., Vinci, A. (eds.) WIVACE 2019. CCIS, vol. 1200, pp. 158–170. Springer, Cham (2020). https://doi.org/10.1007/978-3-030-45016-8_16
23. Sobel, I.: An isotropic 3x3 image gradient operator. Presentation at Stanford A.I. Project 1968 (2014)
24. Stauffer, D., Aharony, A.: Introduction to Percolation Theory. Taylor & Francis (1992)
25. Stokes, G.G.: On the effect of the internal friction of fluids on the motion of pendulums. Trans. Cambridge Philos. Soc. **9**, 8–106 (1851)
26. Szpiro, G.G.: Kepler's Conjecture - How Some of the Greatest Minds in History Helped Solve One of the Oldest Math Problems in the World. John Wiley & Sons, Hoboken (2003)
27. Weyland, M.S., Flumini, D., Schneider, J.J., Füchslin, R.M.: A compiler framework to derive microfluidic platforms for manufacturing hierarchical, compartmentalized structures that maximize yield of chemical reactions. Artif. Life Conf. Proc. **32**, 602–604 (2020). https://doi.org/10.1162/isal_a_00303
28. Wikipedia: Hough transform. https://en.wikipedia.org/wiki/Hough_transform (2021). Accessed 30 Aug 2021
29. Wikipedia: HSV-Farbraum. https://de.wikipedia.org/wiki/HSV-Farbraum (2021). Accessed 30 Aug 2021
30. Wikipedia: Sobel operator. https://en.wikipedia.org/wiki/Sobel_operator (2021). Accessed 30 Aug 2021

The Good, the Bad and the Ugly: Droplet Recognition by a "Shootout"-Heuristics

Hans-Georg Matuttis[1]([✉]), Silvia Holler[2], Federica Casiraghi[2],
Johannes Josef Schneider[3], Alessia Faggian[2], Rudolf Marcel Füchslin[3,4],
and Martin Michael Hanczyc[2,5]

[1] Department of Mechanical and Intelligent Systems Engineering, The University of
Electro-Communications, Chofu Chofugaoka 1-5-1, Tokyo 182-8585, Japan
hg@mce.uec.ac.jp
[2] Laboratory for Artificial Biology, Department of Cellular, Computational and
Integrative Biology, University of Trento, Via Sommarive, 9-38123 Povo, Italy
[3] Institute of Applied Mathematics and Physics, School of Engineering, Zurich
University of Applied Sciences, Technikumstr. 9, 8401 Winterthur, Switzerland
[4] European Centre for Living Technology, S. Marco 2940, 30124 Venice, Italy
[5] Chemical and Biological Engineering, University of New Mexico, MSC01 1120,
Albuquerque, NM 87131-0001, USA

Abstract. We explain how to optimize the image analysis of mixed clusters of red and green droplets in solvents with various degrees of sharpness, brightness, contrast and density. The circular Hough Transform is highly efficient for separated circles with reasonable background contrast, but not for large amounts of partially overlapping shapes, some of them blurred, as in the images of our dense droplet suspensions. We explain why standard approaches for image improvement fail and present a "shootout" approach, where already detected circles are masked, so that the removal of sharp outlines improves the relative optical quality of the remaining droplets. Nevertheless, for intrinsic reasons, there are limits to the accuracy of data which can be obtained on very dense clusters.

Keywords: Image processing · Hough transform · Cluster of droplets in fluids

1 Introduction

It is relatively straightforward to obtain the full information about the geometry of droplet clusters in simulations [10] to verify existing theories with respect to connectivity and packing characteristics [6,8,9], e.g. with applications to chemical reactors of micro-droplets [1,12]. In contrast, experimental verification is difficult due to potentially lost information in the original image. Nevertheless, experimental understanding of the clustering it is indispensable for theoretical predictions concerning the use of DNA technology for DNA-directed self-assembly of oil-in-water emulsion droplets [2]. While the final aim will be to analyse three-dimensional data from confocal microscopy, in this paper, we will

The original version of this chapter was previously published non-open access. A Correction to this chapter is available at https://doi.org/10.1007/978-3-031-23929-8_25

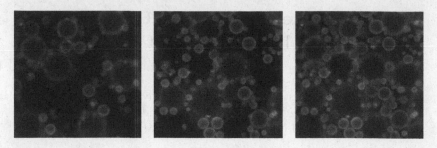

Fig. 1. Image details of droplets with red (left), green (middle) and sum of both red and green (right) color values. (Color figure online)

restrict ourselves to the analysis of two-dimensional conventional images. We will explore the possibilities of extracting as much as possible information about droplet clusters (radii and positions) with the circular Hough transform. It is a standard technique [4,5,7] for recognising shapes in computer vision and image processing (for it's checkered history, see Hart [3]). The circular Hough transform works well for separate circles in two dimensions with good background contrast. This is unfortunately not the case for our graphics: We have several thousand droplets in red and green, partially hidden by each other, with radii from five pixels to several hundred, some out of the focal plane, with varying contrast and sharpness. The red and green droplets have been recorded as tif-RGB-images, with color values either in the red or in the green channel, see Fig. 1. Depending on the choice of color models, interactive programs like Apple's "Digital Color Meter" may indicate mixed colors in the screen display even if the droplets in the original tif image have values only in the red or the green channel. Unfortunately, especially droplets which are larger and out of the focal plain contain different levels of salt and pepper noise (random white, respectively dark pixels). Together with blurred large droplets, there may be small droplets with sharp surface reflections of much higher brightness and scattered reflections inside larger, blurred droplets. In general, very small droplets cannot be discriminated from

Fig. 2. Image details of "good" (left), "bad" (middle) and "ugly" (right) clusters of droplets (sum of red and green color values); from left to right, the number and density of droplets with too different image quality (light and dark, sharp and fuzzy, with and without noise) increases, so that it becomes more difficult to separate individual droplets and determine their radii and positions correctly. (Color figure online)

Fig. 3. Composition of 5 × 5 images (green channel), the original (left) with varying brightness, and the homomorphic colouring (right), with homogeneous, but unfortunately reduced brightness and resolution of details. (Color figure online)

circular scattered reflections, in Fig. 1. From the experiment we have to take the pictures as they are and cannot scan through various focal planes, as is possible with computer tomography or magnetic resonance imagining. All in all we have to deal with three kinds of image qualities as in Fig. 2, which can be catchily expressed as "the Good" (left, reasonable contrast, good separation), "the Bad" (middle, fuzzy contrast, no clear separation) and "the Ugly" (right, deep stacking of droplets to dense clusters with bad contrast, some large droplets only recognisable as "black holes" acting as spacers between smaller droplets). In other words, we have far from perfect input data and when we extract the size and (two-dimensional) position of the droplets, we cannot expect a perfect geometrical representation, only a "best effort" is possible. In this research, we explain how as much information as possible can be extracted for such droplet clusters with methods originally developed for much better resolved circular objects, and which other methods of image processing may help, which may not, and why.

2 Basic Image Processing Approach

2.1 Attempts at Preprocessing to Balance Exposure

Initially, our ambition had been to process the whole of the various sets of 5 × 5 tif-frames, each with resolution of 2048 × 2048. As can be seen in Fig. 3 (left), it turned out that the camera automatically had selected very different individual exposure of each frame, with sufficient time lag between each frame so that lines appeared where the individual tifs were joined together. It is possible to remove the differences in brightness in different regions of images with so-called homomorphic filters [7] with a suitable combination of Fourier transforms and

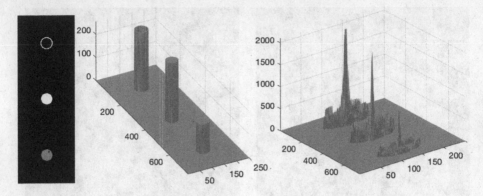

Fig. 4. For an image (left) with a circular ring, a full circle with full (255) and half color saturation (125) - all with the same radius - the corresponding 3D-Graph (middle) as well as the peaks of the circular Hough transform (right).

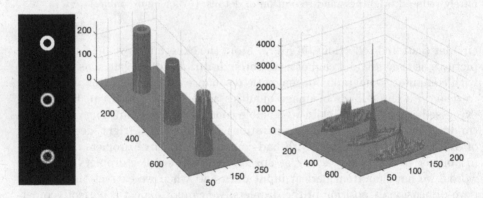

Fig. 5. For an image (left) with a circular ring (above), a circular ring with Gaussian intensity distribution (middle) and the same with 20% salt and pepper noise, the 3D-Graph (middle) as well as the peaks of the circular Hough transform (right).

Gaussian filters. We applied such a filter in Fig. 3 (right), successfully inasmuch as the difference in light and dark regions disappeared. Nevertheless, not only the exposure differences, but also resolution and brightness as a whole were reduced, up to a point where the shape of circles was destroyed. The original aim of analysing whole sets of 5 × 5 frames had to be abandoned, so we decided to focus on the analysis of individual tifs with 2048 × 2048 pixels resolution.

2.2   The Circular Hough Transform

While the standard Hough transform is a method to extract the information about lines (their location and orientation) from an image, the circular Hough transform extracts information about circles (their centers and radii). Simply speaking, the circular Hough transform histograms the possible three-point least-squares fits of a circle for a given center (x, y) and radius r and stores the information in the resulting transform matrix. The range of the radii to be investigated is supplied as input parameter. Because there are more possibilities to fit large than to fit small radii, Hough transforms with larger radii take longer than runs with small radii due to the larger set of the input data. A maximum search on the elements of the transform matrix can then be used to determine the centers and radii of the circles in the original image. Because we will have to deal with droplets which may be imaged as rings or circles with varying shading intensity on the boundary, we will compare in the following the Hough transforms of some circular shapes. In Fig. 4, the circular Hough transform is performed on a monochrome image of 750×250 pixels with (from back to front) a circular ring and a full circle with saturation value 255, and in the foreground there is a circle with saturation value 125. All things being equal, circular rings give Hough transforms with higher peak intensity than full circles, and, as expected, the intensity of the Hough transforms decreases with the color intensity of the circles. In Fig. 5, we have shown the Hough transform (right) for an image of three circular rings (left), with constant intensity (above), with an outline of Gaussian intensity (middle) and with additional 20% salt and pepper noise (front). The peak height for the ring of constant intensity is lower, but wider, while the height of the peaks for the Gaussian outlines is comparable, but the noise reduces the width of the peak.

2.3   The MATLAB Implementation of the Hough Transform

We use the function imfindcircle [11] from MATLAB's image processing toolbox (versions 2017b, respectively 2018b), which adapts the circular Hough transform with additional controls. As input parameters, the preferred range of radii $[r_{min}, r_{max}]$, the polarity ("bright" or "dark" circle to be detected), as well as the sensitivity (between 0 and 1) can be selected. Because most droplet reflections, as in Fig. 1, 2, are not filled circles, but light circular rims with a darker core, the "bright" option for the circle detection must be used: The relative width of the inner "dark" core varies too much to give meaningful radii for the corresponding droplet. MATLAB's imfindcircle is parallelized, and for graphics with 2048×2048 pixels resolution runs in parallel on 2 to 3 cores on a MACBook Pro. For our data, another necessary input is the sensitivity. Here, bigger is not always better, and a sensitivity of 0.9 is reasonable for our images. Beyond a sensitivity of 0.92, imfindcircle starts "seeing things", in particular circles which are not there, and misidentifies irregularities on the outline of droplets, i.e. deviations from an ideal circle, as circular droplets on its own, as can be seen in Fig. 6. Apart from radius and position, imfindcircle outputs also the metric (as it is called in the MATLAB's function description), i.e. the "quality" of the circles. In the case where spurious circles with smaller radii have been inscribed into larger droplets (like droplet 1 and 3 in Fig. 6), these small circles can be eliminated when imfindcircle is run with larger radii and it turns out that several small circles with low metic (quality) are situated inside a larger one with higher

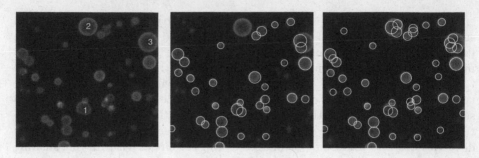

Fig. 6. From left to right: Original picture (left) as well as circles found with imfindcircle with sensitivity of 0.90 (middle) and 0.96 (right) with radii between 6 and 12 pixels. While the droplets 1 and 3 (left) are spuriously resolved with two circles at sensitivity of 0.90 (the outline is not circular) for a sensitivity of 0.96 also droplet 2 gets split into three circles.

metric: In that case, it is clear that the smaller circles were only spurious and can be eliminated.

2.4   Loop over Several Radius-Ranges

For our images, the best performance of imfindcircle is obtained if the input radii are in a range less than a factor of two (all lengths in pixels, input-data in angular brackets in typewrite-font to discriminate them from references): For [r_{min}, r_{max}], [8,12] will give reasonable result when [8,16] may not. Therefore, we run imfindcircle with a sequence of increasing radius ranges and start with the smallest radii, because the smaller droplets are usually the clearer ones. Typical sequences are [5,8], [8,16], [15,26], [27,35], [36,56]. For smaller radii of about 16 pixels, different circles may be recognized for a setting of [8,16] than for a setting of [15,26], for larger radii, the algorithm is more tolerant. With different input radii, the same droplet may be found twice with different radius ranges, e.g. with a radius of 12 for input range [8,16] and with a radius of 16 for an input range of [15,26]: For the smaller input radius, the brightest radius will be detected, for larger radius the largest outer radius may be detected. Therefore, when new sets of positions and radii are obtained with imfindcircle, a loop over all old and new radii must be run to eliminate doublets, concentric circles with different radii but the same center (with an error margin of 1/10 of the larger radius).

3   From Good to Bad Clusters

3.1   Failure of Standard Techniques for Image Improvement

The Hough transform works for the "good" droplet images with reasonable sharpness, brightness, contrast and separation distance from other droplets.

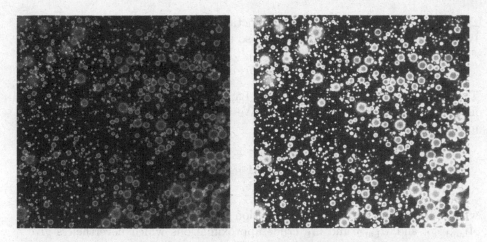

Fig. 7. Original input (left) and result with Laplace filtering using A_{Lap2} (right).

The optical quality of the droplets with small radius and center of mass in the focal plane will always be superior, their signals in the Hough transform always be sharper than for larger droplets or small droplets more distant from this plane. Droplets with less contrast, separation, brightness or sharpness will not be detected with the Hough transform directly, because the corresponding peaks (see Sect. 2.2) are too low. When this happens, everybody who has worked with graphics programs will then try some of the standard methods to improve the image, in particular: brightening, increasing contrast and sharpening contours. All of these methods are implemented in MATLAB, and for interesting reasons turn out to be useless for preprocessing images for use with the Hough transform. Brightening of the input-data will brighten all droplets: The peaks for the sharpest and brightest droplets in the Hough transform will be increased too, and relatively speaking, nothing changes, the droplets with less contrast, brightness etc. will still be outshone. The same is true with increased contrast: In brightened images, the originally brighter droplets will always exhibit the better contrast compared to darker ones. In short, global or unified approaches to improve the picture quality will not help. Sharpening will be discussed in the following section.

3.2   Sharpening Contours with Laplacian Filtering

It is instructive to have a look at what sharpening with Laplacian filtering really accomplishes with a concrete example. Using filters (in digital image processing, i.e. a suitable convolution of an image with a filtering matrix) is a conventional technique to improve image quality. In MATLAB, it is implemented via `imfilter` [11], and the original image and the convolution matrix are input arguments. One of the standard procedures for image sharpening is the Laplace filter. It is derived from the three-point approximation of the Laplacian operator

$\Delta = \partial^2/\partial x^2 + \partial^2/\partial y^2$ for a discrete function $f(x,y)$ on a grid with unit spacing (which takes care of the denominator) so that $\Delta f(x_n, y_n)$ becomes

$$f(x_{n+1}, y_n) + f(x_{n-1}, y_n) + f(x_n, y_{n+1}) + f(x_n, y_{n-1}, y) - 4f(x_n, y_n). \quad (1)$$

The details of the approximation of the Laplacian have much less effect for image filtering than for numerical analysis. We used implementations as convolution masks with the matrices

$$A_{\text{Lap}} = \begin{bmatrix} 0 & -1 & 0 \\ -1 & 4 & -1 \\ 0 & -1 & 0 \end{bmatrix}, \ A_{\text{Lap1}} = \begin{bmatrix} -1 & -1 & -1 \\ -1 & 8 & -1 \\ -1 & -1 & -1 \end{bmatrix}, \ A_{\text{Lap2}} = \begin{bmatrix} 0 & -1 & 0 \\ -1 & 8 & -1 \\ 0 & -1 & 0 \end{bmatrix}, \quad (2)$$

where A_{Lap} is the faithful implementation of the Laplace operator Eq. (1), while A_{Lap1} [5] and A_{Lap2} and are crude approximations which nevertheless give a brighter image. We processed Fig. 7 (left) with A_{Lap2}, so that in the result Fig. 7 (right) the outlines were homogenised and brightened. Unfortunately, this erased other image details, and already bright droplets became brighter, so the relative intensity of the peaks in the Hough transform did not change.

3.3   Failure of More Sophisticated Image Processing Methods

One of the results of Laplace filtering in Fig. 7 (right) was that - while some specific droplets became lighter - in dense clusters, the separation between droplets vanished. To segment pixels between clusters, the watershed [11] transform could be used: Unfortunately, for out amount of data, together with the large size distribution in our images, we were not able to find a suitable separation rule. Also the clearboarder [11] algorithm could not be used, which can remove frayed out borders around separate objects, but not for partially overlapping objects. The temptation to run edge detection algorithms [4,5,7,11] on filtered images like Fig. 7 (right) is overwhelming. However, the results in Fig. 8, where the standard edge detection algorithms have been used, are rather underwhelming: The Sobel [4,5,7]- Roberts- and Prewitt [5] algorithm do not only produce circular shapes, but also any kind of outlines for connected clusters. The contours obtained from the Prewitt algorithm seem to be the most complete, with the

Fig. 8. Edge detection with Sobel (a), Roberts (b), Prewitt (c) and Canny (d) for the filtered data of Fig. 7; shown are the left upper 512×512 of the 2048×2048 pixels.

least amount of gaps. As usual, the Canny [5, 7] algorithm produces the highest detail and most subtle contours. Unfortunately, in case of our ambiguous data that means: There are multiple outlines for the same shape, so the usefulness is rather marginal, compared to the rather more robust Sobel method, which usually produces a single outline. When running `imfindcircle` on the graphics with edge detection in Fig. 8 (a) to (d), the accuracy and resolution was actually reduced compared to the result for the original image.

3.4   The Shootout Method

The recognition of images of mixed optical quality is rather due to the "good" (bright, sharp, well-formed) droplets, because their distinct Hough peaks hide the less-distinct peaks of bad droplets, as in Fig. 9(a). Consequently, it makes sense to mask already recognized droplets. To "shoot out" such detected circles from the image, we have to overwrite them with a suitable color hue. Choosing the wrong color creates spurious contours, i.e. new problems, so we also have to discuss the most practical choice of colors for overwriting droplets. A self-evident choice would be the average or median color of the whole image, as it should - on average - hide a droplet in the average background color. Nevertheless, the medium color of a whole picture with a lot of bright clusters or dark background to overwrite a droplet will create conspicuously bright or dark circles as in Fig. 9(b). So it is not advisable to use color hues which are globally computed for the picture. Each "shoot out" should use a color which is inconspicuous relative to the vicinity of the droplet. That leaves a choice of using either all the data inside a droplet or the values on the circumference. We processed only the values on the circumference, as pixels inside a droplet were sometimes influenced by color deviations. As the diameter found with the Hough transform is very often that of the color maximum at the border and does not include the whole "halo", we used an additional "safety distance" (between 1.1 and 1.2 times the Hough transform-radius) for the shootout. Choosing the darkest pixels on the circumference resulted in spuriously too dark colors as in Fig. 9(c) with artificially sharp contrasts. The best choice was the average color on the circumference, see Fig. 9(d). The replaced circles were most of the time equivalent to the background color and did not induce spurious new circles. As can be seen

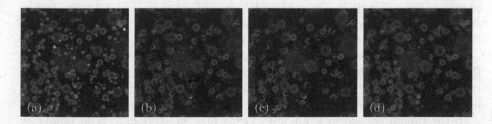

Fig. 9. Original picture (a), shootout of bright dots with average color of the whole frame (b), with darkest color on the circumference (c) and (best choice, to minimize sharp contrasts) with average color of the circumference (d). (Color figure online)

Fig. 10. Droplet recognition for the same image of 2522 green droplets recognized without the shootout procedure (left) and 2719 green droplets recognized with the shootout method (right), with the recognized droplets drawn over the original image. (Color figure online)

in Fig. 10, with the shootout method the number of recognized droplets rises for the green droplets from 2522 to 2719, for the red ones from 1859 to 1999. This may not seem much, but in Fig. 10 (right), one sees that in particular large droplets which got unnoticed without shootout in Fig. 10 (left) can now be recognized. These large droplets with radii between 30 to 60 pixels are practically not detectable with the application of `imfindcircle` alone due to the lack of exposure and contrast.

3.5   Selecting "Ugly" Clusters According to Color Value

As the original purpose of our data evaluation had been the determination of neighbourhoods between red and green droplets, the recognition of larger droplets is essential for a meaningful analysis. Up to here, Fig. 10 looks like a "bad" image which can mostly be dealt with by our shootout method. In reality, the approach has failed, as superpositions hide particularly large droplets with very pale hue, sometimes surrounded by rings of other droplets. These clusters are visible in a certain color range (where 255 corresponds to full saturation): For the green channel between 25 and 80 and for the red channel between 15 and 70, "mega-droplets" appear in white in Fig. 11 which are hardly visible in Fig. 10. Clusters with circles inside could be (more often than not) detected by the circular Hough transform. Some clusters are much larger than the circles inside, or so deformed, as in the left upper corner of Fig. 11(a), that no detection with the Hough transform is possible. Other huge clusters have their boundary and insides decorated with cutouts of other droplets.

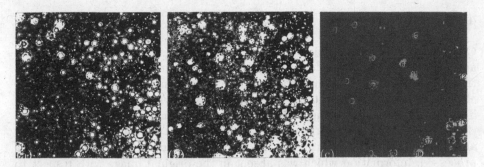

Fig. 11. Very huge droplets for the original image of Fig. 10 for the green channel with color values between 25 and 80 in (a), for the red channel with values between 15 and 70 (b) and semiautomatic recognition of the clusters in (b) shown in (c). (Color figure online)

3.6 The Final Shootout

As in every good Western-themed undertaking, we will finish with a final shootout. The huge droplets could only be recognised by the color hue set by hand, so for green and red droplets, and for different pictures - depending on exposure and density - various values must be selected. This is not practicable for the automatic image analysis we aimed for. Nevertheless, we will now try other methods for cluster recognition, which do not use the Hough transform. For the red channel of the underlying image in Fig. 10, we removed all circular areas with recognized droplets with a shootout of both red and green circles, and binarised the image: All pixels were set to black except red color values between 25 and 48, which were set to white. Over this black-white image, we ran a sequence of MATLAB algorithms for irregular clusters. With `bwareaopen` [11], connected pixel clusters with a total number of 250 pixels or larger are determined, all smaller clusters are removed to erase the pixelated noise. Next, with `imfill` [11], holes inside connected pixel clusters are occluded. At the end, `bwboundaries` [11] is used to label connected pixel clusters. The result is shown in Fig. 11(c) with numbers added for the largest droplets to simplify the discussion. The reconstruction of the droplets 1, 2, 4, 6, 11 and 13 looks mostly intact: The determination of the center and the maximal pixel diameter could give a somehow realistic location of the droplet. Issues exists with droplets 3, 5, 7 and 8 (also droplets 9 and 10, though they are only partially in the image), which appear in several unconnected parts. However, an automatic determination of the connection is not possible: An algorithm which would connect the central pixels of droplets 3, 5, 7 and 8 with their outer rim would also connect areas near droplet 16 and 15, which belong to separate droplets. Finally, the total size of droplets 12, 13, 14, 17 and 18 are unclear, because one cannot determine where the fragments near 13, 14, 17 and 18 belong to. In the case of droplet 12, it is unclear how large the area is which is hidden or cut off by other droplets. Even with all this effort, and selecting the color range by hand, ambiguities about the largest droplets remain.

4   Conclusions and Outlook

Our original aim in processing two-dimensional images of clusters was to obtain automatically an overview over the possible three dimensional contact geometries between droplets from two-dimensional images. Our investigations have clearly shown the limits of this approach: Our shootout-method worked well for droplets with a radius of up to 50 or 60 pixels. It was only possible to extend the recognition to droplets beyond that size by selecting color ranges by hand, and such an approach is futile for the amount of actual data which have to be processed. Information which was absent in the original image with respect to exposure and sharpness cannot be recovered even with the most refined analysis technique. In the future, we will strive for the evaluation of images with higher data content and less noise, from more sophisticated image acquisition methods, in particular from high-resolution fluorescence confocal microscopy.

References

1. Flumini, D., Weyland, M.S., Schneider, J.J., Fellermann, H., Füchslin, R.M.: Towards programmable chemistries. In: Cicirelli, F., Guerrieri, A., Pizzuti, C., Socievole, A., Spezzano, G., Vinci, A. (eds.) Artificial Life and Evolutionary Computation, Wivace 2019, vol. 1200, pp. 145–157 (2020)
2. Hadorn, M., Boenzli, E., Sørensen, K.T., Fellermann, H., Eggenberger Hotz, P., Hanczyc, M.M.: Specific and reversible dna-directed self-assembly of oil-in-water emulsion droplets. Proc. of the Nat. Ac. of Sc. (2012)
3. Hart, P.E.: How the Hough transform was invented. IEEE Signal Process. Mag. **26**(6), 18–22 (2009)
4. Jähne, B.: Digital Image Processing. Springer, Heidelberg (2005)
5. Marques, O.: Practical Image and Video Processing Using MATLAB. Wiley (2011)
6. Müller, A., Schneider, J.J., Schömer, E.: Packing a multidisperse system of hard disks in a circular environment. Phys. Rev. E **79**, 021102 (2009)
7. Parker, J.: Algorithms for Image Processing and Computer Vision. Wiley (1997)
8. Schneider, J.J., Kirkpatrick, S.: Selfish versus unselfish optimization of network creation. J. Stat. Mech. Theory Experiment **2005**(08), P08007 (2005)
9. Schneider, J.J., Müller, A., Schömer, E.: Ultrametricity property of energy landscapes of'multidisperse packing problems. Phys. Rev. E **79**, 031122 (2009)
10. Schneider, J.J., et al.: Influence of the geometry on the agglomeration of a polydisperse binary system of spherical particles. Artificial Life Conference Proceedings, 71 (2021). https://doi.org/10.1162/isal_a_00392
11. The Mathworks: MATLAB helppage. https://uk.mathworks.com/help/index.html (link last followed on July 16, 2021)
12. Weyland, M.S., Flumini, D., Schneider, J.J., Füchslin, R.M.: A compiler framework to derive microfluidic platforms for manufacturing hierarchical, compartmentalized structures that maximize yield of chemical reactions. Artificial Life Conference Proceedings 32, pp. 602–604 (2020)

Exploring the Three-Dimensional Arrangement of Droplets

Johannes Josef Schneider[1(✉)], Mathias Sebastian Weyland[1],
Dandolo Flumini[1], and Rudolf Marcel Füchslin[1,2]

[1] Institute for Applied Mathematics and Physics, Zurich University of Applied Sciences, Technikumstr. 9, 8401 Winterthur, Switzerland
`johannesjosefschneider@googlemail.com`, `{scnj,weyl,flum,furu}@zhaw.ch`
[2] European Centre for Living Technology, Ca' Bottacin, Dorsoduro 3911, Calle Crosera, 30123 Venice, Italy

Abstract. We present some work in progress on the development of a probabilistic chemical compiler, being able to make a plan of how to create a three-dimensional agglomeration of artificial hierarchical cellular constructs. Such programmable discrete units offer a wide variety of technical innovations, like a portable biochemical laboratory being able to produce macromolecular medicine on demand. This paper focuses on one specific issue of developing such a compiler, namely the problem of first studying and then predicting the spatial transition from an originally one-dimensional lineup of droplets into a three-dimensional, almost spherical arrangement, in which the droplets form a network via bilayers connecting them and in which they are contained within some outer hull. The network created by the bilayers allows the droplets to communicate with their neighbors and to exchange chemicals contained within them, thus enabling a complex successive biochemical reaction scheme.

Keywords: Microfluidics · Droplet agglomeration · Ultrametricity

1 Introduction

Over the last decades, huge progress has been made in biochemistry. A large amount of knowledge about the constituents and the processes within a cell has been gathered [1]. Even a new research field, that of "synthetic biology", has evolved [2], in which natural objects like the DNA in cells are purposely altered or replaced in order to achieve some desired outcome, like producing some drug. Still, some questions remain unanswered so far, like one of the basic questions for the origin of life: Which constituent of a cell came first, the RNA [3] or the cell membrane [4]?

In our approach, which we intend to follow within the European Horizon 2020 project *ACDC*, we do not consider fully equipped cells but the most simplified cell-like structures, being droplets comprised of some fluid and surrounded by

Supported by the European Horizon 2020 project *ACDC – Artificial Cells with Distributed Cores to Decipher Protein Function* under project number 824060.
The original version of this chapter was previously published non-open access. A Correction to this chapter is available at https://doi.org/10.1007/978-3-031-23929-8_25

J. J. Schneider et al. (Eds.): WIVACE 2021, CCIS 1722, pp. 63–71, 2022.
https://doi.org/10.1007/978-3-031-23929-8_6

Fig. 1. Sketch of the initial and final states of the spatial rearrangement of droplets.

another fluid. As an additional feature, we also allow droplets being contained within some outer hulls, playing the role membranes have for cells. Droplet generation, especially in the field of microfluidics, has been extensively studied over the past years [5–8] and has become an easy-to-use technology after the introduction of 3D printing technologies [9,10]. A stream of fluid is broken up into droplets within a T-junction or some other antechamber, as the form of spherical droplets is energetically favorable when compared to a continuous stream of fluid. Hereby the applied pressure should be neither too small nor too large, but in a range so that the system is in the so-called dripping regime, in which droplets of fluid are produced in equal time intervals [11]. The size of the droplets can be controlled by the flow rates of the two fluids. In the experiments of our collaborating group in Cardiff, the droplets leave the antechamber, then move lined up in an almost one-dimensional ordering, and enter an expansion chamber while several of them are surrounded by some newly generated hull. Within this capsule, the droplets rearrange themselves in a three-dimensional way [11], as shown schematically in Fig. 1.

This paper is organized at follows: in the next two sections, we describe the steps to be taken before we can start developing a chemical compiler. For this purpose, we first need to simulate the arrangement process and second to compare the resulting configurations. In the last section, we give an outlook to the development of the chemical compiler itself.

2   Simulating the Arrangement Process

In order to study this process and its outcome, we develop a computer simulation. After a short overview of existing and widely used simulation techniques, we present our plan for the generation of a new Monte Carlo Movement Simulation Technique.

2.1   Simulation Techniques for Microfluidic Systems

Over the past years, various approaches for simulating droplets moving in fluids have already been developed, from macroscale approaches, in which not single droplets but only droplet densities are considered, to microscale approaches, in which the state variables of the various droplets are changed gradually and individually. Often the methods had been originally developed for other systems but then adopted for the application to microfluidics.

One macroscale approach is the Lattice Boltzmann method [12,13], with which the time evolution of the density and velocity field of a fluid is simulated on a two-dimensional or three-dimensional lattice. Alternately, collision steps and streaming steps are applied. For the collisions, often the simplified Bhatnagar-Gross-Krook relaxation term [14] is used. While part of the density remains at its current lattice site, other parts are then usually allowed to flow to all sites within a Chebyshev distance of 1, i.e., not only the directly neighboring sites but also the diagonally displaced neighbors are used for flow directions. The huge advantage of this model is that it is very fast and ideally suited for parallel enablement, such that only small amounts of computing time are needed for simulations. The disadvantages are that sometimes lattice artefacts occur and that one has to find out about appropriate rules for flows in various directions. Sometimes two lattices displaced by half a lattice unit in all spatial directions or even more lattices are used instead of one lattice only. A further disadvantage is that this method only considers densities of droplets but not the singular droplets themselves. Thus, this method is not applicable for our investigations, as we need to know about the exact locations and velocities of the various droplets.

On the other hand, the probably most microscopic but also most computer time consuming method is Molecular Dynamics [15,16]. Hereby, the forces between the various particles are considered. The velocities and the locations of the particles are iteratively and simultaneously updated using specific problem-dependent time integrators, which e.g. preserve the total energy of the system. While this method is suited for considering our problem on an atomic or molecular level, we are unable to use it due to the large system size on the one hand and the lack of computing time on the other hand. Thus, we now turn our attention to two types of simulation techniques with an intermediate requirement for computing time, but also with the possibility to simulate the movements of the various droplets in a way that their exact locations and velocities can be determined exactly.

A wide variety of Monte Carlo simulation techniques has been applied to microscopic simulations of discs in two dimensions and spheres in three dimensions for decades, see e.g. [17]. A subclass of these techniques is called the Direct Simulation Monte Carlo technique [18]. It can be applied to study movements in systems for which the mean free path of a particle is of the same size or larger than its representative physical length scale. The method assumes that free movement phases and collision phases can be decoupled over time periods that are smaller than the mean collision time. There are various ways to model collisions [19], some of them seeming to be rather artificial. A widely used modeling of collisions assumes the particles to be point particles. Then a small box is created around a randomly chosen particle. One of the other particles within this box is then randomly chosen for the simulation of a collision process. The velocities of the two particles are taken into account and the rules for a direct collision of these particles if they were point particles are determined. However, some randomness is added to the direction of the relative velocity vector before updating the velocity vectors of the two particles in order to mimick also a non-direct collision of extended particles in a random way. There are also other Monte Carlo approaches like the Griesbauer method [20]. It introduces springs between the various particles, such that we also consider this method not to be applicable for our problem, as the various droplets move entirely independent of each other at the beginning in the experiments, as can be observed in movies generated by Jin Li [21]. Only at a later stage when they are already surrounded by some hull, the droplets gradually settle down, reducing their individual behaviors, and start to move coherently.

A further widely used approach called Dissipative Particle Dynamics [24–26] attempts to relate macroscopic non-Newtonian flow properties of a fluid to its microscopic structure. For the determination of the velocity of the particles, three types of forces are considered which act on a particle. A particle interacts with all other particles within some predefined cut-off distance. There are conservative forces with which the particles interact, then there is a dissipative force, and finally also a random force with zero mean is added. The dissipative and random forces can be chosen in a way that they form a thermostat keeping the mean temperature of the system constant. By choosing the random force between each pair of particles in a way that it acts antisymmetric on both particles as required by Newton's third law, the local momentum of the particles is conserved. Also this technique is well suited for parallel enablement if using spatial domain decomposition. The diameters of the various domains of course need to be much larger than the cut-off distance. However, artefacts due to the spatial decomposition can occur and the geometry of the experiments to be simulated can become rather complex, such that we consider also this method not to be applicable for our problem.

2.2 A New Monte Carlo Movement Simulation Approach

Fig. 2. From left to right: Two droplets in a hull can either stay standalone or touch each other, forming a bilayer to some smaller or larger extent, and either stay almost spherical or lose their spherical shapes.

Summarizing, we intend to create our own simulation technique, with which we want to simulate the experimentally found transition from an originally one-dimensional lineup of droplets into a three-dimensional arrangement. During this rearrangement process, some droplets touching each other will form bilayers [27]. These bilayers can be broken up and reformed, depending on the stability of the bilayers [28]. When bilayers are created, the droplets can lose their spherical shape, as shown in Fig. 2. We will test various ways to simulate the formation, change, and destruction of bilayers and the change of the shape of the cores in a computationally not too expensive way. While the specific spatial setup of an experiment with proposed values for widths and lengths of various parts of the junction can be easily employed also in the Monte Carlo simulation, it is a harder task to find appropriate values for the probabilities for deceleration and acceleration of droplets as well as for bilayer formation and destruction and also for some introduction of random movement. These values depend on various experimental parameters, like pressure and viscosity, and also on the various radius values of the droplets. We intend to adjust the parameters for the Monte Carlo simulation in a way that the resulting configurations reflect the three-dimensional arrangements of droplets as found in experiments.

3 Comparing Resulting Configurations

Fig. 3. Top: Two resulting three-dimensional arrangements of droplets filled with various chemicals. Bottom: Corresponding bilayer networks between the droplets. We made these bilayer networks visible by reducing the size of the droplets and printing a connecting edge between a pair of neighboring droplets if they have formed a bilayer.

After this first part of our objective has been achieved, we have a closer look at the resulting arrangements. As the experiments performed by Jin Li have already shown [21], there is not the one and only resulting three-dimensional packing of droplets. Instead, various arrangements are possible, as depicted in Fig. 3. However, when looking closely at the resulting configurations, one finds that they are not entirely random but often rather similar to each other and that maybe even some configurations can be considered as part of a group of configurations having several properties in common. Whether such groups of configurations have an entirely identical backbone [29–31] in common or whether they share some properties with some larger probabilities, as found for dense packings of multidisperse systems of hard discs [32] will be seen. We also need to question the influence of the excess of polydispersity of the radius values.

Fig. 4. Left: Overlap matrix with a structure dominated by iterated replica symmetry breaking. Right: corresponding ultrametric tree.

Complex systems often exhibit the property of ultrametricity in configuration space [22, 23]. A standard metric d has to obey to the triangle inequality

$$d(i, j) \leq d(i, k) + d(k, j), \tag{1}$$

with $d(i, j)$ denoting the distance from node i to node j, i.e., a direct connection cannot be longer than a detour via a third node k. For an ultrametric, this inequality is replaced by the ultrametricity condition

$$d(i, j) \leq \max\{d(i, k), d(k, j)\}. \tag{2}$$

If permuting the nodes i, j, and k, one finds that this condition is fulfilled if the nodes are placed on the edges of equilateral triangles or isosceles triangles with short base. A distance between two configurations can be defined using an overlap measure between the configurations. The larger the overlap is, the smaller is the distance. After the application of an appropriate permutation of these configurations, which can e.g. be found with an optimization technique leading to a clustered ordering of configurations [33], the overlap matrix can exhibit a structure as schematically depicted in the left graphics of Fig. 4. In this example, one finds that the overall set of configurations is split in two large groups of configurations. The overlap values of the configurations to other configurations in the same group are larger than those to configurations in the other group. Each group can then be split in four subgroups in this example, which in turn exhibit even larger overlap values within each subgroup. The subgroups are then split again. In statistical physics, one speaks of iterated replica symmetry breaking if such a behavior is observed. Ideally, this replica symmetry breaking property corresponds to ultrametricity, i.e., one can derive also another representation by generating an ultrametric tree. The right graphic in Fig. 4 shows such a tree. At the root, the tree splits into two branches, which in turn split into four subbranches each. These subbranches then split again into three subbranches each. The various configurations then form the leaves of the tree on the right side.

As the property of ultrametricity was also found for a related hard disc packing problem [34], we expect it will also turn up for this problem. As already mentioned, ultrametricity is related to iterated replica symmetry breaking and the possibility to generate ultrametric trees. For their generation, we will use the neighbor-joining method, which is a standard tool to reconstruct phylogenetic trees [35, 36], as well as finding a clustered ordering of configurations [33].

4   Final Steps Towards a Probabilistic Biochemical Compiler

If we have achieved this second part of our objective of understanding and predicting the outcome of an experimental setup, i.e., when the various groups of three-dimensional arrangements of droplets have been generated, we will be able to create a probabilistic chemical compiler in the final stage of this project. We aim at creating plans for e.g. a step-wise generation of some desired macromolecules, which are gradually constructed from smaller units, contained in the various droplets, with the successive chemical reactions enabled via the bilayers formed between neighboring droplets. Thus, the compiler has

– to determine bilayer networks with which the desired reaction chains leading e.g. to the macromolecules we want to produce can be performed and
– to design and to govern the experiment leading to such a bilayer network.

Such a compiler has been exemplarily already developed for one specific molecule [37]. In this project, this compiler has to be generalized and also made probabilistic because of the various possible outcomes in the rearrangement process.

References

1. Alberts, B., Johnson, A., Lewis, J., Morgan, D., Raff, M., Roberts, K., Walter, P.: Molecular Biology of The Cell, Garland Science, 6th edn. Taylor & Francis, New York (2014)

2. Gibson, D.G., Hutchison, C.A., III., Smith, H.O., Venter, J.C. (eds.): Synthetic Biology - Tools for Engineering Biological Systems. Cold Spring Harbor Laboratory Press, Cold Spring Harbor, New York (2017)

3. Higgs, P.G., Lehman, N.: The RNA World: molecular cooperation at the origins of life. Nat. Rev. Genet. **16**, 7–17 (2015)

4. Lancet, D., Segré, D., Kahana, A.: Twenty years of "lipid world": a fertile partnership with David Deamer. Life **9**, 77 (2019)

5. Eggers, J., Villermaux, E.: Physics of liquid jets. Rep. Prog. Phys. **71**, 036601 (2008)

6. Link, D.R., Anna, S.L., Weitz, D.A., Stone, H.A.: Geometrically mediated breakup of drops in microfluidic devices. Phys. Rev. Lett. **92**, 054503 (2004)

7. Garstecki, P., Fuerstman, M.J., Stone, H.A., Whitesides, G.M.: Formation of droplets and bubbles in a microfluidic T-junction - scaling and mechanism of breakup. Lab Chip **6**, 437–446 (2006)

8. Guillot, P., Colin, A., Ajdari, A.: Stability of a jet in confined pressure-driven biphasic flows at low Reynolds number in various geometries. Phys. Rev. E **78**, 016307 (2008)

9. Au, A.K., Huynh, W., Horowitz, L.F., Folch, A.: 3D-printed microfluidics. Angew. Chem. Int. Ed. **55**, 3862–3881 (2016)

10. Tasoglu, S., Folch, A. (eds.): 3D Printed Microfluidic Devices. MDPI, Basel (2018)

11. Li, J., Barrow, D.A.: A new droplet-forming fluidic junction for the generation of highly compartmentalised capsules. Lab Chip **17**, 2873–2881 (2017)

12. Chen, S., Doolen, G.D.: Lattice Boltzmann method for fluid flows. Annu. Rev. Fluid Dyn. **30**, 329–364 (1998)

13. McNamara, G., Garcia, A., Alder, B.: A hydrodynamically correct thermal lattice Boltzmann model. J. Stat. Phys. **87**, 1111–1121 (1997)

14. Bhatnagar, P.L., Gross, E.P., Krook, M.: A model for collision processes in gases. I. Small amplitude processes in charged and neutral one-component systems. Phys. Rev. **94**, 511–525 (1954)

15. Griebel, M., Knapek, S., Zumbusch, G.: Numerical Simulation in Molecular Dynamics. Springer, Heidelberg (2007)

16. Bou-Rabee, N.: Time integrators for molecular dynamics. Entropy **16**, 138–162 (2014)

17. Metropolis, N., Rosenbluth, A.W., Rosenbluth, M.N., Teller, A.H., Teller, E.: Equation of state calculations by fast computing machines. J. Chem. Phys. **21**, 1087–1092 (1953)

18. Bird, G.A.: Approach to translational equilibrium in a rigid sphere gas. Phys. Fluids **6**, 1518–1519 (1963)

19. Roohi, E., Stefanov, S.: Collision partner selection schemes in DSMC: From micro/nano flows to hypersonic flows. Phys. Rep. **656**, 1–38 (2016)

20. Griesbauer, J., Seeger, H., Wixforth, A., Schneider, M.F.: Method for the Monte Carlo Simulation of Lipid Monolayers including Lipid Movement. https://arxiv.org/pdf/1012.4973.pdf (2010), and references therein

21. Li, J.: Private communication (2019)

22. Parisi, G.: Infinite number of order parameters for spin-glasses. Phys. Rev. Lett. **43**, 1754–1756 (1979)

23. Parisi, G.: A sequence of approximate solutions to the S-K model for spin glasses. J. Phys. A **13**, L-115 (1980)

24. Hoogerbrugge, P.J., Koelman, J.M.V.A.: Simulating microscopic hydrodynamic phenomena with dissipative particle dynamics. Europhys. Lett. **19**, 155–160 (1992)

25. Koelman, J.M.V.A., Hoogerbrugge, P.J.: Dynamic simulations of hard-sphere suspensions under steady shear. Europhys. Lett. **21**, 363–368 (1993)
26. Español, P., Warren, P.: Statistical mechanics of dissipative particle dynamics. Europhys. Lett. **30**, 191–196 (1995)
27. Mruetusatorn, P., Boreyko, J.B., Venkatesan, G.A., Sarles, S.A., Hayes, D.G., Collier, C.P.: Dynamic morphologies of microscale droplet interface bilayers. Soft Matter **10**, 2530–2538 (2014)
28. Guiselin, B., Law, J.O., Chakrabarti, B., Kusumaatmaja, H.: Dynamic morphologies and stability of droplet interface bilayers. Phys. Rev. Lett. **120**, 238001 (2018)
29. Schneider, J., Froschhammer, C., Morgenstern, I., Husslein, T., Singer, J.M.: Searching for backbones - an efficient parallel algorithm for the traveling salesman problem. Comp. Phys. Comm. **96**, 173–188 (1996)
30. Schneider, J.: Searching for Backbones - a high-performance parallel algorithm for solving combinatorial optimization problems. Futur. Gener. Comput. Syst. **19**, 121–131 (2003)
31. Schneider, J.J.: Searching for backbones - an efficient parallel algorithm for finding groundstates in spin glass models. In: Tokuyama, M., Oppenheim, I.: 3rd International Symposium on Slow Dynamics in Complex Systems, Sendai, Japan. AIP Conference Proceedings 708, pp. 426–429 (2004)
32. Müller, A., Schneider, J.J., Schömer, E.: Packing a multidisperse system of hard disks in a circular environment. Phys. Rev. E **79**, 021102 (2009)
33. Schneider, J.J., Bukur, T., Krause, A.: Traveling salesman problem with clustering. J. Stat. Phys. **141**, 767–784 (2010)
34. Schneider, J.J., Müller, A., Schömer, E.: Ultrametricity property of energy landscapes of multidisperse packing problems. Phys. Rev. E **79**, 031122 (2009)
35. Saitou, N., Nei, M.: The neighbor-joining method: a new method for reconstructing phylogenetic trees. Mol. Biol. Evol. **4**, 406–425 (1987)
36. Studier, J.A., Keppler, K.J.: A note on the neighbor-joining algorithm of Saitou and Nei. Mol. Biol. Evol. **5**, 729–731 (1988)
37. Weyland, M.S., Fellermann, H., Hadorn, M., Sorek, D., Lancet, D., Rasmussen, S., Füchslin, R.M.: The MATCHIT automaton: exploiting compartmentalization for the synthesis of branched polymers. Comput. Math. Methods Med. **2013**, 467428 (2013)

Geometric Restrictions to the Agglomeration of Spherical Particles

Johannes Josef Schneider[1](✉)(iD), David Anthony Barrow[2](iD), Jin Li[2](iD),
Mathias Sebastian Weyland[1](iD), Dandolo Flumini[1](iD), Peter Eggenberger Hotz[1],
and Rudolf Marcel Füchslin[1,3](iD)

[1] Institute of Applied Mathematics and Physics, School of Engineering, Zurich
University of Applied Sciences, Technikumstr. 9, 8401 Winterthur, Switzerland
johannesjosefschneider@googlemail.com,
{scnj,weyl,flum,eggg,furu}@zhaw.ch

[2] Laboratory for Microfluidics and Soft Matter Microengineering, Cardiff School of
Engineering, Cardiff University, Queen's Buildings, 14-17 The Parade, Cardiff, Wales
CF24 3AA, United Kingdom
{Barrow,LiJ40}@cardiff.ac.uk

[3] European Centre for Living Technology, Ca' Bottacin, Dorsoduro 3911,
Calle Crosera, 30123 Venice, Italy
https://www.zhaw.ch/en/about-us/person/scnj/

Abstract. Within the scope of the European Horizon 2020 project
*ACDC – Artificial Cells with Distributed Cores to Decipher Protein
Function*, we aim at the development of a chemical compiler governing the three-dimensional arrangement of droplets, which are filled with
various chemicals. Neighboring droplets form bilayers containing pores
through which chemicals can move from one droplet to its neighbors.
When achieving a desired three-dimensional configuration of droplets,
we can thus enable gradual biochemical reaction schemes for various
purposes, e.g., for the production of some desired macromolecules for
pharmaceutical purposes. In this paper, we focus on geometric restrictions to possible arrangements of droplets. We present analytic results
for the buttercup problem and a heuristic optimization method for the
kissing number problem, which we then apply to find (quasi) optimum
values for a bidisperse kissing number problem, in which the center sphere
exhibits a larger radius.

Keywords: Agglomeration · Droplet · Buttercup problem · Kissing
number · Threshold accepting

1 Introduction

Over the last decades, huge progress has been made in discovering the properties
and functionalities of the various constituents of cells [2]. In the new field of syn-

This work has been partially financially supported by the European Horizon 2020
project *ACDC – Artificial Cells with Distributed Cores to Decipher Protein Function*
under project number 824060.

The original version of this chapter was previously published non-open access. A Correction to this chapter is available at https://doi.org/10.1007/978-3-031-23929-8_25

J. J. Schneider et al. (Eds.): WIVACE 2021, CCIS 1722, pp. 72–84, 2022.
https://doi.org/10.1007/978-3-031-23929-8_7

thetic biology [7], this knowledge is used to produce some desired macromolecules by altering or replacing the DNA. Within the scope of the European Horizon 2020 project *ACDC – Artificial Cells with Distributed Cores to Decipher Protein Function*, we follow a different approach: We use droplets containing some fluid and being surrounded by another fluid as simplified models of cell-like structures. Droplets touching each other can form bilayers with pores through which chemicals can be exchanged. Thus, an agglomeration of droplets filled with various chemicals can form a bilayer network, allowing a gradual chemical reaction process. Our aim is to create a specific agglomeration of droplets in order to generate a desired gradual chemical reaction process for some purpose, like the generation of some desired macromolecule [17,18]. In order to reach this goal, we need to study the dynamics and outcome of agglomeration processes [20], but also to develop a chemical compiler [6,23], creating a plan of how to design an experiment leading to an appropriate agglomeration for a bilayer network for the desired gradual reaction processes.

Besides, we are also interested in the origin of life [15]. One of the open questions in this field is the huge difference between the time needed for a totally random evolutionary process, starting from the primordial soup and finally leading to life of larger complexity, up to human beings, and the age of the earth, which is too small to allow such an entirely random evolutionary process performed at a constant rate, such that faster rates in the past have to be assumed [10]. But if we replace the primordial soup at least partially with an agglomeration of droplets [27], the production of the first complex molecules of life might become likelier and thus faster, as the compartimentalization of semi-stable molecules within small droplets instead of within the overall primordial soup might stabilize these educts [11], thus increase the probability for the production of such complex molecules [15], and in turn shorten the time for some initial steps of the evolutionary process.

For both of these problems we want to deal with, the basic question arises which molecules could be created by an agglomeration of droplets and the resulting bilayer network. And vice versa, one could ask which agglomeration was needed to produce a specific desired molecule. But here an additional problem arises: Even if a bilayer network for the creation of that molecule can be determined theoretically, still the question remains whether this bilayer network can indeed be realized with an agglomeration of droplets filled with various chemicals. Geometric restrictions may form the most prominent obstacles towards the realizations of such bilayer networks.

In this paper, we restrict ourselves to the case that the droplets can be modeled as hard spheres, i.e., we consider the case of large surface tensions, such that the droplets have in general a spherical shape, and of small bilayers, such that the droplets only touch slightly each other and stay spherical if a bilayer is formed. Furthermore, we mostly restrict ourselves to the case of all spheres exhibiting the same radius value R. In the following sections, we present two examples of geometric restrictions of great importance for our underlying problems, show how one of these problems can be solved analytically, develop a heuristic optimization algorithm for finding (quasi) optimum solutions for the other problem, and present results obtained with that algorithm.

2   The Buttercup Problem

2.1   Description

Fig. 1. Solutions for the buttercup problem for $N = 2$, $N = 3$, $N = 4$, and $N = 5$, (top row from left to right) unit spheres on a ring (printed in red), with all of them touching one further unit sphere (printed in blue), with a glossy sphere encapsulating them and touching all of them. For $N = 6$, such a configuration is already impossible for a finite radius of the surrounding hull, such that we print here the configuration with the smallest hull sphere encapsulating a ring of 6 unit spheres (bottom). (Color figure online)

The first problem we want to consider here has been named the buttercup problem (see Fig. 1): N unit spheres (We use the term unit sphere throughout this paper for a sphere with radius 1. But all results shown for these unit spheres can be easily transferred to other systems with spheres having an other common radius value R.) shall be placed around an additional $N + 1$th center unit sphere in their midst in the way

- that each of the N unit spheres touch the center unit sphere in their midst,
- that the N unit spheres form a ring, i.e., each of them touches only its left and its right neighbor besides touching the center sphere and no further connections are formed, and
- that the whole configuration is placed within a surrounding hull sphere of radius $\mathcal{R}$, with $\mathcal{R}$ being chosen in a way that the overall configuration is stabilized without further ado, i.e., the surrounding hull touches each of the N spheres on the ring and the center sphere.

From these conditions, it follows that the midpoints of the N unit spheres on a ring lie in one plane. This structure is reminiscent to a buttercup flower with its petals and center. It is important for our problem of generating specific agglomerations for some desired purposes, as the center sphere can serve as a control sphere which governs e.g. the opening and closing of pores in the bilayers between the petal spheres.

2.2   Analytic Solution

Now we have to determine the configurations for various N. Without restriction, we place the midpoints of the N petal spheres in the xy-plane on a circle with radius ϱ, such that the midpoint of the petal sphere No. k has the coordinates $(\varrho\cos(2\pi k/N), \varrho\sin(2\pi k/N), 0)$. For symmetry reasons, the $N+1$th center sphere has the coordinates $(0, 0, z_{N+1})$ and the midpoint of the surrounding hull, which has a radius value of $\mathcal{R}$, lies at $(0, 0, h)$. First we determine the radius ϱ with the condition

$$(\varrho\cos(2\pi/N) - \varrho)^2 + (\varrho\sin(2\pi/N))^2 = 2^2, \tag{1}$$

which resolves to

$$\varrho = \sqrt{\frac{2}{1 - \cos(2\pi/N)}}. \tag{2}$$

There are of course also some more elegant ways to determine ϱ for some specific values of N. For example, for the case $N = 5$, one can make use of the properties of the golden cut and finds $\varrho = \frac{1}{5}\sqrt{50 + 10\sqrt{5}}$. In the second step, we determine z_{N+1} with the condition

$$z_{N+1}^2 = 4 - \varrho^2. \tag{3}$$

So far, the locations of the $N + 1$ spheres have already been determined. Now we have to choose h and $\mathcal{R}$ in a way that the outer hull touches all $N + 1$ spheres, which is only the case if there is a common distance value $\mathcal{D}$ between the midpoint of the surrounding hull and the midpoints of the $N + 1$ spheres. For determining the two parameters h and $\mathcal{R}$, we need two conditions:

$$\begin{aligned} h^2 + \varrho^2 &= \mathcal{D}^2 \\ z_{N+1} - h &= \mathcal{D} \end{aligned} \tag{4}$$

Please note that there are two cases: If the midpoints of the $N + 1$th sphere and of the surrounding hull lie on opposite sides of the xy-plane, then h is negative, as in the case $N = 5$. If they lie on the same side, then h is positive, as for $N \leq 3$. By eliminating $\mathcal{D}$ from Eq. (4), we get

$$h = \frac{z_{N+1}^2 - \varrho^2}{2z_{N+1}}. \tag{5}$$

The radius $\mathcal{R}$ can then easily be determined as

$$\mathcal{R} = z_{N+1} - h + 1. \tag{6}$$

Table 1 provides the parameters of the buttercup problem for $N = 2$ to $N = 6$. Please note that the configuration for $N = 6$ in Fig. 1 violates the condition that the surrounding hull must touch all spheres, including the center

Table 1. Approximate numeric values of the parameters (see text) of the solutions to the buttercup problem for various numbers N of unit spheres in a ring. If the spheres have a common radius value R different from 1, each of the numbers provided has to be multiplied by R.

N	ϱ	z_{N+1}	h	$\mathcal{R}$
2	1	1.732	0.577	2.155
3	1.155	1.633	0.408	2.225
4	1.414	1.414	0	2.414
5	1.701	1.051	−0.851	2.902
6	2	0	−∞	∞

sphere. This is only possible in the limit $\mathcal{R} \to \infty$. Instead the picture for $N = 6$ depicts the case of the smallest surrounding sphere. But this configuration is not stable, as the center sphere can move freely in the z-direction. The configurations for $N = 2, 3$, and 4 are identical with the densest packings of 3, 4, and 5 spheres in a sphere [26]. In the configuration for $N = 4$, even a further unit sphere can be added symmetric to the $N + 1$th sphere at the other side of the xy-plane without overlaps and without the need to enlarge $\mathcal{R}$. Furthermore note that no buttercup configurations can be created for $N > 6$ (see also the configurations for the related problem of packing circles in a circle in [24]), if not at least one of the conditions mentioned above is abandoned or the radius of the center sphere enlarged.

3   The Kissing Number Problem

3.1   Description

Fig. 2. The kissing number problem in three dimensions

The second problem we want to study in this paper is the kissing number problem [3,4]. It can be stated as follows: find the maximum number N of unit spheres touching a further unit sphere in their midst without overlaps in D dimensions [16]. This problem, which has been studied for centuries, can be trivially solved in one and two dimensions. In two dimensions, $N = 6$ unit disks can be placed around a center disk. In one dimension, the kissing number is given by $N = 2$. But in three dimensions, the exact value remained unknown for a long time. Newton and Gregory debated about this problem, with Newton claiming that the kissing number in three dimensions was $N = 12$, whereas Gregory believed it was 13. Only in the 1950s, it was proved that Newton had been right [22]. A configuration of an arrangement of spheres for this kissing number problem in three dimensions is shown in Fig. 2, with the center sphere printed in blue and the surrounding spheres printed in red. A glassy orb is drawn around the configuration to better show that the surrounding spheres indeed touch the center sphere in their midst. The kissing number problem is also considered in higher dimensions, but only for some dimensions like $D = 4$ ($N = 24$), $D = 8$ ($N = 240$), and $D = 24$ ($N = 196560$), exact values of N have been determined. For most dimensions, only lower and upper bounds to the kissing number are known. A list of currently known values can be found in [25]. They indicate that the kissing number N grows exponentially with the dimension D. In the next step, we want to develop an optimization algorithm for the kissing number problem.

3.2   A Heuristic Optimization Algorithm for the Kissing Number Problem

Optimization algorithms can be divided in two classes, namely exact mathematical algorithms, which return one solution and guarantee that this solution is indeed optimal, and heuristic methods, which create one or several configurations of good quality, hopefuly being even (quasi) optimum. These heuristic methods can be subdivided in two subclasses, one of them being construction heuristics, which start at a tabula rasa and gradually create a solution for the overall problem, and the other one being iterative improvement heuristics, which usually start at a randomly chosen configuration and iteratively apply changes to the configuration in order to gradually increase the quality [21]. Mostly, one follows the local search [1] path and applies only small changes which do not alter a configuration very much. The simplest improvement heuristic is the greedy algorithm: starting from a randomly chosen initial configuration σ_0, it performs a series of moves $\sigma_i \rightarrow \sigma_{i+1}$ by successively applying small randomly chosen changes to the configuration. A move is accepted with probability

$$p(\sigma_i \rightarrow \sigma_{i+1}) = \begin{cases} 1 & \text{if } \Delta\mathcal{H} \leq 0 \\ 0 & \text{otherwise} \end{cases} \tag{7}$$

with the difference $\Delta\mathcal{H} = \mathcal{H}(\sigma_{i+1}) - \mathcal{H}(\sigma_i)$ between the cost function values of the current configuration σ_i and the tentative new configuration σ_{i+1}. In case of rejection, one sets $\sigma_{i+1} := \sigma_i$.

As the greedy algorithm does not accept any deteriorations, it often gets soon stuck at local minima of mediocre quality. In contrast, simulated annealing [9], which uses the Metropolis criterion [12]

$$p(\sigma_i \rightarrow \sigma_{i+1}) = \begin{cases} 1 & \text{if } \Delta\mathcal{H} \leq 0 \\ \exp(-\Delta\mathcal{H}/T) & \text{otherwise} \end{cases} \tag{8}$$

with the temperature T as acceptance criterion, also accepts deteriorations with some probability, which is the smaller the larger the deterioration $\Delta\mathcal{H}$ and the smaller the temperature T is. Its deterministic variant [13], which is also called threshold accepting [5], uses the transition probability

$$p(\sigma_i \rightarrow \sigma_{i+1}) = \begin{cases} 1 & \text{if } \Delta\mathcal{H} \leq T \\ 0 & \text{otherwise} \end{cases} \tag{9}$$

and thus accepts all improvements and all deteriorations up to a proposed threshold value T. During the optimization run, T is gradually decreased from a large initial value, at which the system to be optimized is in a high-energetic unordered regime, towards a small vanishing value, at which the system gradually freezes in a low-energetic ordered solution.

When applying threshold accepting to a specific optimization problem like the kissing number problem, one thus has to define

- a routine for creating a random initial configuration,
- one or several move routines, which alter a configuration slightly, and
- a function for measuring the cost function value of a configuration.

In our optimization apprach, we use threshold accepting, as its acceptance criterion is better at avoiding small overlaps at the end of a simulation than the Metropolis criterion. We work with a fixed number N of spheres, which shall be placed in a way that all of them touch the center sphere and that they do not overlap with each other. If a feasible configuration meeting these constraints is found for a specific value of N, we increase N by 1 and perform a new optimization run with this incremented value of N. We iterate this approach until no feasible configuration can be found anymore for a specific $N_{\max}$, such that the heuristic algorithm returns that the kissing number has the value $N_{\max} - 1$.

For the initialization process, we could randomly place the N spheres anywhere. However, we want to restrict the search space to configurations in which the midpoints of the N spheres lie on a virtual sphere with radius 2, as this condition must hold for any solution of the problem. For all initial coordinates $x_i(j), j = 1, \ldots, D$ of sphere i, we first randomly select a value from the interval $[-2, 2]$, such that the midpoints of the spheres are already placed in a D-dimensional cube of side length 4, which is centered around the origin, at the beginning. Then we calculate the distance

$$r_i = \sqrt{\sum_{j=1}^{D} x_i(j)^2} \tag{10}$$

of the midpoint of sphere i to the origin and renormalize the coordinates of the midpoint of sphere i according to

$$x_i(j) := 2 \times x_i(j)/r_i, \tag{11}$$

thus placing the midpoint of sphere i on the virtual sphere of radius 2. Therefore, the condition that each of the N spheres has to touch the sphere at the center is already fulfilled.

Furthermore, we implemented three move routines in which one of the N spheres is randomly chosen and transferred to a tentative new location:

- In the first of these three move routines, the new location of the chosen sphere is determined as in the initialization process, i.e., first each of its coordinates are randomly chosen in a way that its midpoint lies in a cube of side length 4 centered around the origin, then they are renormalized using Eq. 11, i.e., in the way that the midpoint lies again on the virtual sphere of radius 2. Thus, the midpoint sphere can lie at an entirely new location on this virtual sphere.
- In the second move routine, the tentative new location is created by adding randomly chosen values from the interval $[-1, 1]$ to the current coordinates of sphere i, such that the midpoint of sphere i in the tentative new configuration first lies in a cube of side length 2 centered around the current position. Then the coordinates are renormalized as above, such that the midpoint lies on the virtual sphere of radius 2 again. We chose this maximum value of 1 for shifting a coordinate, as this maximum value proved to provide superior results for a formerly studied multidisperse disk packing problem [14, 19].
- The third move routine is almost identical to the second move routine, except that the interval from which the values to be added to the coordinates are randomly chosen is not $[-1, 1]$ but $[-T, T]$. With decreasing threshold value T, the amount by which the coordinates can be changed at a maximum decreases during the optimization run, thus increasing the likelihood for randomly selecting small changes with larger acceptance probability.

Each of these three move routines ensures that the N spheres touch the center sphere as required, however, configurations can contain overlaps between pairs of these surrounding spheres.

This brings us to the last point of our considerations of how to apply threshold accepting to the kissing number problem. For the application of an optimization algorithm like threshold accepting, a cost function $\mathcal{H}$ is needed which is to be minimized. However, there is no such cost function for the kissing number problem in our approach with a fixed number N of surrounding spheres a priori, one can only state whether a configuration is feasible because all constraints are fulfilled or whether it is not feasible because at least one pair of surrounding spheres overlaps. However, during the optimization run in the search of a feasible configuration, we can allow to accept non-feasible configurations as well, but introduce penalties for the overlaps between pairs of spheres, which we sum up and thus get a cost function value. Therefore, we can define the cost function $\mathcal{H}$ as

$$\mathcal{H}(\sigma) = \sum_{\substack{i,j=1 \\ i<j}}^{N} \Delta(i,j)\Theta(\Delta(i,j)) \tag{12}$$

with

$$\Delta(i,j) = 2 - \sqrt{\sum_{k=1}^{D}(x_i(k) - x_j(k))^2} \tag{13}$$

and the Heaviside step function

$$\Theta(a) = \begin{cases} 1 & \text{if } a \geq 0 \\ 0 & \text{otherwise.} \end{cases} \tag{14}$$

When a tentative new configuration is put to the decision whether to accept or reject it, not its overall cost function value has to be calculated. When determining $\Delta\mathcal{H}$ for an algorithm using local search moves as in our approach, it is much faster only to sum up the differences between those addends which have changed between the current and the tentative new configuration, i.e., one has to check only for the sphere which is to be displaced whether there are overlaps at its current position and whether there are at the tentative new position.

The threshold T is gradually decreased from an initial value of 10 by a cooling factor of 0.999. In each of the 10000 temperature steps, one million MCS are performed. An optimization run can be prematurely stopped when a feasible configuration without overlaps is reached.

3.3   Computational Results for the Kissing Number Problem

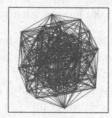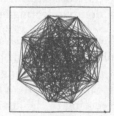

Fig. 3. Two solutions each of our optimization program for the kissing number problem in $D = 4$ dimensions with $N = 24$ surrounding spheres (left) and in $D = 6$ dimensions with $N = 72$ surrounding spheres (right): Only the midpoints of the N surrounding spheres are displayed at $(x_i(1), x_i(2))$. If surrounding spheres touch each other, their midpoints are connected with edges. The central sphere is omitted for visibility reasons.

We applied the threshold accepting algorithm as described above and got the solution shown in Fig. 2 for the kissing number problem in three dimensions. But the kissing number problem does not provide much of a challenge in three dimensions, as the available space on the surface of the central sphere is almost sufficient for a thirteenth sphere. In order to show the effectiveness of our heuristic algorithm, we thus applied the algorithm to the kissing number problem also in higher dimensions. In several optimization runs, we achieved feasible solutions for the optimum number of $N = 24$ surrounding spheres in $D = 4$ dimensions and for $N = 72$ surrounding spheres in $D = 6$ dimensions, four of which are shown in Fig. 3. We also tried to find a feasible solution with $N = 73$ surrounding spheres in $D = 6$ dimensions, but were unsuccessful. The best solutions we achieved were configurations with a cost function value of 2, i.e., in these solutions, the 73rd surrounding sphere is placed exactly where the 72nd surrounding sphere lies. Although having no exact proof by using threshold accepting, we thus assume that the value of $N = 72$ is not only the lower bound to the true kissing number in six dimensions but that it truely is the kissing number in six dimensions.

3.4 Computational Results for a Bidisperse Kissing Number Problem

Fig. 4. Solutions for a bidisperse kissing number problem with a center sphere with larger radius R: $N = 28$ unit spheres can be placed around this larger center sphere without overlaps for $R = 2$, $N = 52$ for $R = 3$, $N = 83$ for $R = 4$, and $N = 120$ for $R = 5$ (from left to right).

Finally, we apply our optimization algorithm to a bidisperse variant of the kissing number problem in which the center sphere has no longer a radius of $R = 1$ but a larger radius value, while the surrounding spheres are still unit spheres, i.e., still have a radius of 1. For this problem, we have to alter our optimization algorithm simply in the way that we have to ensure that the midpoints of the surrounding spheres lie on a virtual sphere of radius $R + 1$, therefore Eq. (11) has to be replaced by

$$x_i(j) := (R + 1) \times x_i(j)/r_i. \tag{15}$$

Otherwise, the optimization algorithm remains unchanged. We consider this altered problem only in three dimensions. Figure 4 displays the best solutions found for various radius values. The configurations are drawn as in Fig. 2. The kissing number N increases with increasing radius R of the center sphere. Due to the small problem sizes studied, we cannot generally state according to which laws N will furthermore increase if we move on to even larger values of R. However, if considering that the surface of the center sphere is given by $4\pi R^2$, we can construct an upper bound function to the kissing number N, which increases $\propto R^2$, such that the kissing number itself cannot increase stronger than with $\propto R^2$.

4   Conclusion and Outlook

In this paper, we considered two problems which impose restrictions to our aim of generating bilayer networks in agglomerations of droplets for the purpose of producing specific macromolecules. In the buttercup problem, spheres have to be placed like the petals of a buttercup around a center sphere while fulfilling some constraints. In the kissing number problem, the largest number of spheres being able to touch a further sphere in their midst without overlaps has to be determined. We provided analytic results for the first problem and obtained computational results from a heuristic optimization algorithm we had developed for the second problem, achieving feasible configurations with optimum kissing numbers in 3 and 4 dimensions and with the lower bound in 6 dimensions. Furthermore, we provided results for a newly introduced bidisperse variant of the kissing number problem. We will continue studying these problems and their variants with additional constraints. We will also consider other such problems related to our work, like the problem of sticky spheres [8], in which agglomerations of spheres have to be found in a way that the number of bilayers between them is maximum.

Acknowledgments. JJS would like to thank Sebiha Şahin for a former collaboration on kissing numbers at the Johannes Gutenberg University of Mainz, Germany, in 2010 [16] and Hermann Stamm-Wilbrandt for an introduction to the "packing circles in a circle" problem [24] at the former IBM Scientific Center Heidelberg, Germany in 1995. Furthermore, he would like to thank his former teachers Rudolf Piller, Walter Mielach, and Josef Graminger at the Ludwigsgymnasium Straubing, Germany for torturing him with a lot of homework in mathematics. Finally, he is grateful to Ingo Morgenstern (formerly at University of Regensburg, Germany), Gunter Dueck, Tobias Scheuer, Gerhard Schrimpf (formerly at IBM Heidelberg), Scott Kirkpatrick (The Hebrew University of Jerusalem, Israel), and Elmar Schömer (University of Mainz) for fruitful discussions about optimization algorithms.

References

1. Aarts, E., Lenstra, J.K.: Local Search in Combinatorial Optimization. Princeton University Press, Princeton and Oxford (2003)
2. Alberts, B., Johnson, A., Lewis, J., Morgan, D., Raff, M., Roberts, K., Walter, P.: Molecular Biology of The Cell, 6th edn. Taylor & Francis, New York, Garland Science (2014)
3. Brass, P., Moser, W., Pach, J.: Research problems in discrete geometry. Springer (2005)
4. Conway, J.H., Sloane, N.J.: Sphere Packings, Lattices and Groups, 3rd edn. Springer, New York (1999)

5. Dueck, G., Scheuer, T.: Threshold accepting: a general purpose optimization algorithm appearing superior to simulated annealing. J. Comput. Phys. **90**(1), 161–175 (1990)
6. Flumini, D., Weyland, M.S., Schneider, J.J., Fellermann, H., Füchslin, R.M.: Towards programmable chemistries. In: Cicirelli, F., Guerrieri, A., Pizzuti, C., Socievole, A., Spezzano, G., Vinci, A. (eds.) WIVACE 2019. CCIS, vol. 1200, pp. 145–157. Springer, Cham (2020). https://doi.org/10.1007/978-3-030-45016-8_15
7. Gibson, D., Hutchison, C., III., Smith, H., Venter, J.E.: Synthetic Biology - Tools for Engineering Biological Systems. Cold Spring Harbor Laboratory Press, Cold Spring Harbor, New York (2017)
8. Hayes, B.: The science of sticky spheres - on the strange attraction of spheres that like to stick together. Am. Sci. **100**, 442 (2012)
9. Kirkpatrick, S., Gelatt, C.D., Vecchi, M.P.: Optimization by simulated annealing. Science **220**(4598), 671–680 (1983)
10. Knoll, A.H., Nowak, M.A.: The timetable of evolution. Sci. Adv. **3**(5), e1603076 (2017)
11. Mann, S.: Wie entsteht leben: Ein altes problem gebiert neue chemie. Angew. Chem. **125**, 166–173 (2013)
12. Metropolis, N., Rosenbluth, A.W., Rosenbluth, M.N., Teller, A.H., Teller, E.: Equation of state calculations by fast computing machines. J. Chem. Phys. **21**(6), 1087–1092 (1953)
13. Moscato, P., Fontanari, J.: Stochastic versus deterministic update in simulated annealing. Phys. Lett. A **146**(4), 204–208 (1990)
14. Müller, A., Schneider, J.J., Schömer, E.: Packing a multidisperse system of hard disks in a circular environment. Phys. Rev. E **79**, 021102 (2009)
15. Oparin, A.: The Origin of Life on the Earth, 3rd edn. Academic Press, New York (1957)
16. Şahin, S.: Physikalische Optimierung ausgewählter Packprobleme. Master's thesis, Johannes Gutenberg University of Mainz, Germany (2010)
17. Schneider, J.J., Weyland, M.S., Flumini, D., Füchslin, R.M.: Investigating three-dimensional arrangements of droplets. In: Cicirelli, F., Guerrieri, A., Pizzuti, C., Socievole, A., Spezzano, G., Vinci, A. (eds.) Artificial Life and Evolutionary Computation, Wivace 2019, Communications in Computer and Information Science, vol. 1200, pp. 171–184 (2020)
18. Schneider, J.J., Weyland, M.S., Flumini, D., Matuttis, H.-G., Morgenstern, I., Füchslin, R.M.: Studying and simulating the three-dimensional arrangement of droplets. In: Cicirelli, F., Guerrieri, A., Pizzuti, C., Socievole, A., Spezzano, G., Vinci, A. (eds.) WIVACE 2019. CCIS, vol. 1200, pp. 158–170. Springer, Cham (2020). https://doi.org/10.1007/978-3-030-45016-8_16
19. Schneider, J.J., Müller, A., Schömer, E.: Ultrametricity property of energy landscapes of multidisperse packing problems. Phys. Rev. E **79**, 031122 (2009)
20. Schneider, J.J., et al.: Influence of the geometry on the agglomeration of a polydisperse binary system of spherical particles. vol. ALIFE 2021: The 2021 Conference on Artificial Life (2021). https://doi.org/10.1162/isal_a_00392
21. Schneider, J.J., Kirkpatrick, S.: Stochastic Optimization. Springer, Berlin Heidelberg, Germany (2006)
22. Szpiro, G.G.: Kepler's Conjecture: How Some of the Greatest Minds in History Helped Solve One of the Oldest Math Problems in the World. John Wiley & Sons (2003)

23. Weyland, M.S., Flumini, D., Schneider, J.J., Füchslin, R.M.: A compiler framework to derive microfluidic platforms for manufacturing hierarchical, compartmentalized structures that maximize yield of chemical reactions. In: Artificial Life Conference Proceedings 32, pp. 602–604 (2020). https://doi.org/10.1162/isal_a_00303

24. Wikipedia: Circle packing in a circle (2021). https://en.wikipedia.org/wiki/Circle_packing_in_a_circle. Accessed 31 July 2021

25. Wikipedia: Kissing number (2021). https://en.wikipedia.org/wiki/Kissing_number. Accessed 31 July 2021

26. Wikipedia: Sphere packing in a sphere (2021). https://en.wikipedia.org/wiki/Sphere_packing_in_a_sphere. Accessed 31 July 2021

27. Zwicker, D., Seyboldt, R., Weber, C.A., Hyman, A.A., Jülicher, F.: Growth and division of active droplets provides a model for protocells. Nat. Phys. **13**, 408–413 (2017)

Effectiveness of Dynamic Load Balancing in Parallel Execution of a Subsurface Flow Cellular Automata Model

Andrea Giordano[1]([✉]) [iD], Donato D'Ambrosio[2] [iD], Alessio De Rango[3] [iD], Luca Furnari[3] [iD], Rocco Rongo[2] [iD], Alfonso Senatore[3] [iD], Giuseppe Mendicino[3] [iD], and William Spataro[2] [iD]

[1] ICAR-CNR, Rende, Cosenza, Italy
giordano@icar.cnr.it
[2] Department of Mathematics and Computer Science, University of Calabria, Arcavacata, Italy
{d.dambrosio,rongo,spataro}@unical.it
[3] Department of Environmental Engineering, University of Calabria, Arcavacata, Italy
{alessio.derango,luca.furnari,alfonso.senatore,
giuseppe.mendicino}@unical.it

Abstract. In this paper, a subsurface flow Cellular Automata (CA) model, namely the XCA-Flow model, is considered with the aim of optimizing its parallel execution by means of a purposely tailored dynamic load balancing technique. Indeed, a suitable distribution of computational load over different processing elements is particular relevant in the case of parallel execution of CA, where the domain space is partitioned in regions assigned to the parallel computing nodes. In addition, the XCA-Flow model can exhibit very unbalanced distribution of the water flow, and this unbalanced condition also might change during the simulation advancement. As a consequence, a Dynamic Load Balancing technique can be suitably utilized in order to achieve an optimal resource utilization thus reducing the overall execution time. First tests implemented using the MPI technology have demonstrated an appreciable reduction of execution times in comparison with the not-balanced parallel version.

Keywords: Cellular automata · Parallel computing · Load balancing · Subsurface flow

1 Introduction

Efficient solutions for large data-intensive computational problems resort to advanced Parallel Computing methodologies. High Performance Computing (HPC) adopts numerical simulations as a tool for solving complex equation systems which rule the dynamics of complex systems as swarm intelligence algorithms [4], lava flow [8], fire spreading [2], or unmanned aerial vehicles simulation [6]. Classical approaches based on calculus, usually represented by Partial

© The Author(s), under exclusive license to Springer Nature Switzerland AG 2022
J. J. Schneider et al. (Eds.): WIVACE 2021, CCIS 1722, pp. 85–96, 2022.
https://doi.org/10.1007/978-3-031-23929-8_8

Differential Equations - PDEs, often fail to solve these kinds of equations analytically, making a numerical computer-based methodology mandatory in case of solutions for real situations. In this context, numerical methods such as Finite Volume Method (FEM) and Cellular Automata (CA), are widely adopted for the simulation of natural phenomena. Notwithstanding their simple definition, CA [40] are universal computational models [38], widely adopted in both theoretical studies [42], and applications [1,19,25,30], and can produce extremely complex behavior at a macroscopic level, making them suitable for the simulation of complex natural processes. Even if local laws that determine the dynamics of the system are known, the system's global behavior can be hard to be predicted, giving rise to what is called *emergent behaviour*. Many CA applications and models can be found in the literature, regarding both theoretical and scientific fields (e.g., [1,9,15,17–19,25–28,30,32,33,35,37,42]).

In this paper, we consider a hydrological CA model, namely the XCA-Flow subsurface flow model [13], with the aim of improving performances in parallel execution. Thanks to their implicit parallel nature, CA models can be efficiently parallelized on a set of computing nodes to scale and speed up their execution (e.g., [3,10,14]). Typically, the CA space is partitioned by adopting a data-parallel scheme in regions, where the execution of each of them is assigned to a different computing node. Indeed, several studies regarding performance improvement in CA parallelization can be found in the literature, such as in [11,15,22,23].

As expected, applying parallel computing undoubtedly has shown its effectiveness for the improvement of CA execution. However, the expected benefits that are obtained can sometimes be counterbalanced by overheads that can arise due to the parallelization process itself [5,24], as parallel tasks inevitably need to exchange data during computation. For this reason, apart from an unavoidable communication burden, even idle times can be produced by the synchronization process which is required for proper inter-process interaction [7]. In this case, synchronization issues can deeply affect the overall performances when a load imbalance execution occurs. In such a context, processing elements that have a minor assigned workload must wait before they can continue execution for the more overloaded ones. Therefore, idle times are added to the overall execution time when an inter-process interaction is required. As a consequence, *load balancing* techniques can offer a valuable solution aiming to reduce the overhead related to synchronization.

Load Balancing (LB) [41] is generally referred to as the process to properly partition computation between processing elements with the aim of making their overall processing more efficient, and thus reducing the overall computing time. The parallel system configuration and the nature of interactions that occur among the concurrent tasks usually determine whether a *static* or a *dynamic* LB technique can be suitable. In static LB the tasks are distributed to the processing elements before execution, while in dynamic LB the workload is dynamically distributed during the execution of the algorithm.

In this paper, a load balancing approach is applied to the parallel execution of the XCA-Flow CA model [13]. Due to its nature, a *dynamic* LB was adopted since the affected area by the subsurface water flow changes dynamically during the simulation advancement. The carried out experiments confirm the advantage of adopting the load balancing technique to the execution of the XCA-Flow CA model in terms of speedup improvement.

The paper is organized as follows. In Sect. 2 the XCA-Flow Subsurface flow CA model is briefly introduced; subsequently, in Sect. 3 a schema for the efficient parallelization of the XCA-Flow CA model is presented, while in Sect. 4, the adopted LB technique is detailed. In Sect. 5, experimental results are given confirming the goodness of the LB approach in speed up enhancement. Eventually, conclusions and future developments (Sect. 6) conclude the paper.

2 The XCA-Flow Subsurface Flow CA Model

The XCA-Flow subsurface water flow model is based on a time-explicit finite differences scheme already presented in [31], [12] and [20]. In particular, the unknown state variable is the total hydraulic head h [m], given by the sum of the pressure head ψ [m] and the elevation z [m]. This variable is updated according to a volumetric mass balance equation. The XCA-Flow is modelled as a cellular automata and, the simulated subsurface test case, drawing inspiration by [36] and [29], consisting of a 16 m wide and 13 m deep rectangular domain, was discretized with a 0.1 m resolution regular mesh, thus producing a 160×130 CA cells. On all domain boundaries, Neumann's no-flow conditions were fixed, except along the first 3.4 m starting from the left of the upper border, where a constant rainfall rate of 0.02 m d^{-1} is considered. The initial conditions consist of a fixed total hydraulic head of -7.34 m all over the computational domain.

The execution of XCA-Flow model, following the typical CA execution scheme, consists of a step-by-step evaluation of the so-called *transition function* for each CA cell. The transition function applied on a given cell takes as input the state of the cell itself together with the state of its *neighbours* cells and produces the new cell state as output.

The implementation of the XCA-Flow transition function takes into consideration a random, but realistic, generated heterogeneous field of saturated hydraulic conductivity K_s [m s^{-1}]. Since K_s can easily range on many orders of magnitude, due to the extremely complex structure of the soil, the logarithmic transform of the variable was generated. The constitutive relationship between pressure head and volumetric moisture content was given by the Van Genuchten's equations [39] in the extended form proposed in [34] to ensure continuity between the unsaturated and the saturated zone. To this aim, a continuity parameter ψ_0 was fixed equal to 0.001 m and a related specific storage S_s equal to 10^{-6} m^{-1}. The other hydrological parameters were fixed as follows: residual volumetric moisture content $\theta_r = 0.0955$ [m^3 m^{-3}], saturated volumetric moisture content $\theta_s = 0.348$ [m^3 m^{-3}], and finally the two Van Genuchten's equations parameters $\alpha = 0.03473$ [m $^{-1}$] and n = 1.729 [-]. Figure 1 illustrates the final step of the

48-days simulation of the variation of volumetric moisture content $\Delta\theta$ [-] over the computation domain.

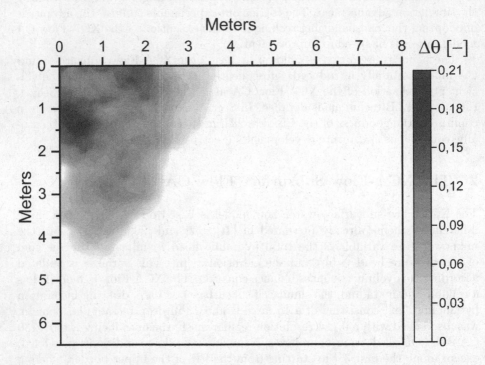

Fig. 1. The soil moisture content variation $\Delta\theta$ [−] at the final step of the 48-days reference simulation.

3   Parallel Execution of XCA-Flow in Distributed Memory Architectures

The execution of the XCA-Flow CA model on a single computer is quite straightforward and concerns the evaluation of the transition function over all the cells of the cellular space, which can be evaluated sequentially or in a parallel fashion. Typically, the parallel execution of CA consists in partitioning the space and assigning each subspace (or region) to a specific processing element [16], which is in charge of applying the transition function to all the cells belonging to the specific region. The state of a CA is typically stored in two matrices: the *read* and the *write* matrix. The input of the transition function is taken by values of the read matrix and its output written on the write matrix. At the end of each step, the two matrices are swapped so a new CA step can begin.

Since the computation of a transition function of a cell is based on the states of the neighbour cells, information is required from adjacent computing nodes in order to evaluate the transition function for cells located at the edge of a region. Thus, at each computing step, the states of border cells (or *halo* cells) need to be exchanged among adjacent nodes in order to keep the parallel execution consistent.

Eventually, the parallel execution of a CA in distributed memory is summarized by the following pseudo-code Algorithm 1 which is executed on each processing element.

Algorithm 1: CA execution

```
1 while !StopCriterion()              // Loop until CA stop criterion met
2 do
3    SendBorderToNeighbours() // Send halo borders to neighbour nodes
4    ReceiveBorderFromNeighbours() // Receive halo borders from
         neighbour nodes
5    ComputeTransitionFunction() // Read from read matrix and write
         to write matrix
6    SwapReadWriteMatrices() // Swap read and write matrices
7    step ← step + 1                               // Next CA step
```

---

4 Load Balancing of XCA-Flow CA Execution

The XCA-Flow CA model is likely to give rise to an imbalance execution since a simulation can initially develop within a (usually small) sub-region of the surface area, to further expand in the remaining subsurface area. As a consequence, a LB procedure can be useful to improve computational performances.

The adopted LB algorithm, detailed in [21], consists in dynamically computing the optimal workload that each processing element has to take into account for achieving uniform execution times over a simulation. The computation load is exchanged among computing nodes at a predefined rate of CA time steps.

In this preliminary application, the load balancing is actually achieved by means of columns exchange among nodes by considering the whole 2D domain partitioned along the X dimension. In particular, each node exchanges columns with its neighbour nodes in a parallel fashion by exploiting the same communication channel already in use for the halo exchange. At each CA step, the execution times experienced by the computing nodes are retrieved and stored in each node. When the LB phase has to be executed, each node sends its step time to a specific "master" node, which is in charge of establishing a suitable columns exchanges that nodes must perform in order to achieve a balanced workload.

Algorithm 2: CA execution with Dynamic Load Balancing

```
1  while !StopCriterion() do
2  │  if IsLoadBalancingPhase() then
3  │  │  SendToMaster({myStepTime, myWorkload})
4  │  │  LBInfo = ReceiveFromMaster()
5  │  │  if IamLBMaster then
6  │  │  │  {stepTime, workload}[]=ReceiveFromNodes()
7  │  │  │  LBInfo[] = LoadBalance({stepTime, workload}[])
8  │  │  └  SendToNodes(LBInfo[])
9  │  └  ExchangeWorkload(LBInfo)
   │     // Back to normal CA execution
10 │  SendBorderToNeighbours()
11 │  ReceiveBorderFromNeighbours()
12 │  ComputeTransitionFunction()
13 └  step ← step + 1
```

The modified CA loop containing the code implementing the load balancing is illustrated in pseudo-code Algorithm 2. When `IsLoadBalancingPhase()` is `true` (line 2), i.e., at a predefined CA step rate, the LB phase is actually executed. If this is the case, at line 3, each node sends the experienced step time and its workload, i.e., the number of columns, to the master node (`SendToMaster()`). Afterwards, each node waits, from master (`ReceiveFromMaster()`, line 4), information (`LBInfo`) about the exchange of workload that the node has to perform to achieve load balancing. Finally, at line 9 the actual workload exchange takes place.

The master node behaviour takes place at lines 6–8. At line 6, the master receives information about the state of nodes (i.e., step time and workload) and determines, at line 7 (`LoadBalance()`), the workload' exchange *schema* for all the nodes that is, at line 8, broadcasted to all the nodes.

The core of this LB procedure relies on the function `LoadBalance()`, which determines *how* the workload has to be redistributed among nodes and will be specified in the next subsection.

4.1   The Determination of the Optimal Load Balance

As previously stated, the Master node determines the optimal workload exchange in the `LoadBalance()` function. Figure 2 graphically shows the load exchange problem referred to a set of N regions with dimensions equal to S_i, $0 \leq i < N$, columns. The nodes step times experienced during the last computational step (i.e., before the load balancing step) are indicated as T_i, $0 \leq i < N$. Let us define Δx_i, $0 \leq i < N-1$ as the exchanges of columns that take place between nodes i and $i+1$.

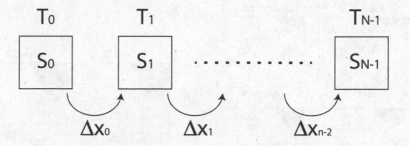

Fig. 2. Load balancing scheme for a twodimensional CA partitioned and balanced along the x dimension. For each node i, S_i represents the size (i.e., the number of columns in this case), T_i the experienced step time and Δx_i the workload exchange between nodes i and $i+1$.

The load balancing problem consists in finding the values of Δx_i that, starting from a current system state $\langle S_i, T_i \rangle$, produce the new system state $\langle S_i', T_i' \rangle$ where $T_0' = T_1' = \ldots = T_{N-1}'$. The formulas for the optimal solution for the above-mentioned problem are taken from [21] and, specifically, from the expressions (3), (4), (5), and (6) of that work where these formulas are derived.

5   Experimental Results

A series of experiments were executed for evaluating the performance of the proposed LB algorithm on the model XCA-Flow described in Sect. 2. The experiments were conducted on a scientific workstation running Archlinux (Linux kernel 5.4) and equipped with two Intel 8 core (16 threads) Xeon E5- 2650 sockets (for a total of 32 threads), and a total of 32 GB of RAM. The opensource C++ OpenMPI 4 was used for message passing among MPI processes, by measuring the speedup values referred to a number of processes from 1 to 16. The simulation was carried out for 40000 CA steps, corresponding to a 48-days period. The load balancing phase is executed every 7777 steps, corresponding to a total number of five LB applications during the simulation. A first set of experiments were carried out in order to show how the XCA-Flow model execution evolves in the parallel scenario when the load balancing procedure is applied. The Fig. 3 illustrates the five LB phases, by showing the step just before and after the LB application. The vertical black lines delimit the region of each of the 16 MPI processes. As noted, for each of the five LB phases, the new region sizes (i.e., "After LB" in the figure) that are determined by the algorithm quasi-optimally partition the CA domain[1] to the previous computational load (i.e., "Before LB" in the figure), determined by the wet front infiltration represented by the blue area, which dynamically changes during the simulation.

---

[1] The not completely optimal behaviour is due to the non-negligible computation load of the "stationary" cells, i.e. those ones that are not involved by the wet water front.

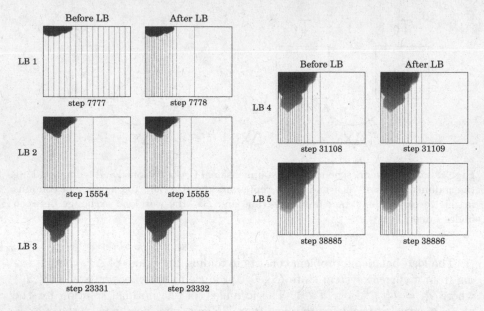

Fig. 3. The five LB phases in the 16 MPI processes parallel execution. The blue coloring represents the wet water front, the vertical black lines are the limits of each partitioning region. For each LB phase, the state of the parallel execution is portrayed just before the LB phase and after the LB phase, respectively. (Color figure online)

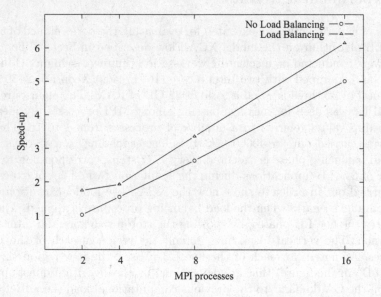

Fig. 4. Speedup comparison between balanced and non-balanced execution vs number of MPI processes.

The second set of experiments concerns the evaluation of the parallel performances by adopting the speedup indicator, evaluated as the ratio between the elapsed times of a single node run and the elapsed time needed for completing the parallel execution. In particular, Fig. 4 illustrates the comparison between the non-balanced and load balanced execution by varying the number of MPI processes from 1 to 16. The figure clearly shows the advantage of adopting the load balancing procedure, which outperforms the normal execution in all cases, by reaching an improvement ranging from 22 % (16 MPI processes) to 69 % (2 MPI processes).

6 Conclusions

This paper presents a dynamic load balancing application for reducing overall execution times in parallel executions of the XCA-Flow CA hydrological model on distributed memory architectures. Specifically, the algorithm executes load balancing among processors to reduce processor timings at regular intervals, based on a procedure that computes the optimal distribution load exchange among adjacent processes. Preliminary experiments, devoted to assessing the advantage of the dynamically load-balanced version with respect to the non-balanced one, resulted in good improvements for the considered testbed, with improvements up to 69%. Future work will regard the adoption of other CA partitionings, such as two-dimensional space partitioning of regions for the case of 2D CA, where the LB phase can take place also along rows of the CA regions. Moreover, additional experiments will be carried out to compute the most beneficial LB parameters, besides considering other LB strategies. For instance, a criterion can consider the LB phase only when elapsed times between nodes are significant instead of prefixed step intervals, and other heuristics to compute the new node workload. Finally, LB experiments will consider the XCA-Flow model development, coupled with a surface routing component, to simulate real case studies at the basin scale.

References

1. Aidun, C., Clausen, J.: Lattice-Boltzmann method for complex flows. Annu. Rev. Fluid Mech. **42**, 439–472 (2010)
2. Arca, B., Ghisu, T., Trunfio, G.A.: Gpu-accelerated multi-objective optimization of fuel treatments for mitigating wildfire hazard. J. Comput. Sci. **11**, 258–268 (2015). https://doi.org/10.1016/j.jocs.2015.08.009
3. Cannataro, M., Di Gregorio, S., Rongo, R., Spataro, W., Spezzano, G., Talia, D.: A parallel cellular automata environment on multicomputers for computational science. Parallel Comput. **21**(5), 803–823 (1995)
4. Cicirelli, F., Forestiero, A., Giordano, A., Mastroianni, C.: Transparent and efficient parallelization of swarm algorithms. ACM Trans. Autonomous Adaptive Syst. (TAAS) **11**(2), 14 (2016)
5. Cicirelli, F., Forestiero, A., Giordano, A., Mastroianni, C.: Parallelization of space-aware applications: modeling and performance analysis. J. Netw. Comput. Appl. **122**, 115–127 (2018)

6. Cicirelli, F., Furfaro, A., Giordano, A., Nigro, L.: An agent infrastructure for distributed simulations over hla and a case study using unmanned aerial vehicles. In: 40th Annual Simulation Symposium, 2007, ANSS 2007, pp. 231–238. IEEE (2007)
7. Cicirelli, F., Giordano, A., Mastroianni, C.: Analysis of global and local synchronization in parallel computing. IEEE Trans. Parallel Distrib. Syst. **32**(5), 988–1000 (2020)
8. Crisci, G.M., et al.: Predicting the impact of lava flows at mount etna, Italy. J. Geophys. Res. Solid Earth **115**(B4) (2010)
9. Crisci, G.M., Gregorio, S.D., Rongo, R., Spataro, W.: Pyr: a cellular automata model for pyroclastic flows and application to the 1991 mt. pinatubo eruption. Future Gener. Comput. Syst. **21**(7), 1019–1032 (2005)
10. D'Ambrosio, D., De Rango, A., Oliverio, M., Spataro, D., Spataro, W., Rongo, R., Mendicino, G., Senatore, A.: The open computing abstraction layer for parallel complex systems modeling on many-core systems. J. Parallel Distribut. Comput. **121**, 53–70 (2018)
11. D'Ambrosio, D., Filippone, G., Marocco, D., Rongo, R., Spataro, W.: Efficient application of gpgpu for lava flow hazard mapping. J. Supercomput. **65**(2), 630–644 (2013)
12. De Rango, A., et al.: Opencal system extension and application to the three-dimensional richards equation for unsaturated flow. Computers and Mathematics with Applications (2020)
13. De Rango, A., Furnari, L., Senatore, A., D'Ambrosio, D., Straface, S., Mendicino, G.: Massive simulations on gpgpus of subsurface flow on heterogeneous soils, pp. 249–252 (2021). https://doi.org/10.1109/PDP52278.2021.00047, cited By 0
14. De Rango, A., Spataro, D., Spataro, W., D'Ambrosio, D.: A first multi-gpu/multi-node implementation of the open computing abstraction layer. J. Comput. Sci. **32**, 115–124 (2019). https://doi.org/10.1016/j.jocs.2018.09.012, cited By 10
15. Di Gregorio, S., Filippone, G., Spataro, W., Trunfio, G.: Accelerating wildfire susceptibility mapping through gpgpu. J. Parallel Distrib. Comput. **73**(8), 1183–1194 (2013)
16. D'Ambrosio, D., Filippone, G., Rongo, R., Spataro, W., Trunfio, G.A.: Cellular automata and gpgpu: an application to lava flow modeling. Int. J. Grid High Perform. Comput. **4**(3), 30–47 (2012)
17. Filippone, G., D'Ambrosio, D., Marocco, D., Spataro, W.: Morphological coevolution for fluid dynamical-related risk mitigation. ACM Trans. Model. Comput. Simul. (TOMACS) **26**(3), 18 (2016)
18. Folino, G.: Cellar: a high level cellular programming language with regions, pp. 259–266, February 2000
19. Frish, U., Hasslacher, B., Pomeau, Y.: Lattice gas automata for the Navier- Stokes equation. Phys. Rev. Lett. **56**(14), 1505–1508 (1986)
20. Furnari, L., Senatore, A., De Rango, A., De Biase, M., Straface, S., Mendicino, G.: Asynchronous cellular automata subsurface flow simulations in two- and three-dimensional heterogeneous soils. Adv. Water Resour. **153** (2021). https://doi.org/10.1016/j.advwatres.2021.103952, cited By 0
21. Giordano, A., De Rango, A., Rongo, R., D'Ambrosio, D., Spataro, W.: Dynamic load balancing in parallel execution of cellular automata. IEEE Trans. Parallel Distrib. Syst. **32**(2), 470–484 (2021). https://doi.org/10.1109/TPDS.2020.3025102, cited By 1
22. Giordano, A., D'Ambrosio, D., De Rango, A., Portaro, A., Spataro, W., Rongo, R.: Exploiting distributed discrete-event simulation techniques for parallel execution

of cellular automata. In: Cicirelli, F., Guerrieri, A., Pizzuti, C., Socievole, A., Spezzano, G., Vinci, A. (eds.) WIVACE 2019. CCIS, vol. 1200, pp. 66–77. Springer, Cham (2020). https://doi.org/10.1007/978-3-030-45016-8_8

23. Giordano, A., De Rango, A., D'Ambrosio, D., Rongo, R., Spataro, W.: Strategies for parallel execution of cellular automata in distributed memory architectures. In: 2019 27th Euromicro International Conference on Parallel, Distributed and Network-Based Processing (PDP), pp. 406–413. IEEE (2019)

24. Grama, A.Y., Gupta, A., Kumar, V.: Isoefficiency: measuring the scalability of parallel algorithms and architectures. IEEE Parallel Distrib. Technol. Syst. Appl. **1**(3), 12–21 (1993)

25. Higuera, F., Jimenez, J.: Boltzmann approach to lattice gas simulations. Europhys. Lett. **9**(7), 663–668 (1989)

26. Langton, C.: Computation at the edge of caos: phase transition and emergent computation. Physica D **42**, 12–37 (1990)

27. LucÃ, F., D'Ambrosio, D., Robustelli, G., Rongo, R., Spataro, W.: Integrating geomorphology, statistic and numerical simulations for landslide invasion hazard scenarios mapping: An example in the sorrento peninsula (italy). Comput. Geosci. **67**(1811), 163–172 (2014)

28. Macri, M., Rango, A., Spataro, D., D'Ambrosio, D., Spataro, W.: Efficient lava flows simulations with opencl: a preliminary application for civil defence purposes, pp. 328–335 (2015). https://doi.org/10.1109/3PGCIC.2015.107,cited By 4

29. McCord, J.T., Goodrich, M.T.: Benchmark testing and independent verification of the VS2DT computer code. Technical report, Sandia National Labs (1994)

30. McNamara, G., Zanetti, G.: Use of the Boltzmann equation to simulate lattice-gas automata. Phys. Rev. Lett. **61**, 2332–2335 (1988)

31. Mendicino, G., Senatore, A., Spezzano, G., Straface, S.: Three-dimensional unsaturated flow modeling using cellular automata. Water Resour. Res. **42**(11), 2332–2335 (2006)

32. Ninagawa, S.: Dynamics of universal computation and 1/f noise in elementary cellular automata. Chaos, Solitons Fractals **70**(1), 42–48 (2015)

33. Ntinas, V.G., Moutafis, B.E., Trunfio, G.A., Sirakoulis, G.C.: Parallel fuzzy cellular automata for data-driven simulation of wildfire spreading. J. Comput. Sci. **21**, 469–485 (2017)

34. Paniconi, C., Aldama, A.A., Wood, E.F.: Numerical evaluation of iterative and noniterative methods for the solution of the nonlinear richards equation. Water Resour. Res. **27**(6), 1147–1163 (1991)

35. Renc, P., Pecak, T., De Rango, A., Spataro, W., Mendicino, G., Was, J.: Towards efficient gpgpu cellular automata model implementation using persistent active cells. J. Comput. Sci. **59**, 101538 (2022). https://doi.org/10.1016/j.jocs.2021.101538

36. Smyth, J., Yabusaki, S., Gee, G.: Infiltration evaluation methodology-letter report 3: Selected tests of infiltration using two-dimensional numerical models. Pacific Northwest Laboratory, Richland, WA (1989)

37. Spataro, D., D'Ambrosio, D., Filippone, G., Rongo, R., Spataro, W., Marocco, D.: The new sciara-fv3 numerical model and acceleration by gpgpu strategies. Int. J. High Perform. Comput. Appl. **31**(2), 163–176 (2017)

38. Thatcher, J.W.: Universality in the von neumann cellular model. In: Burks, A.W. (ed.) Essays on Cellular Automata, chap. 5, pp. 132–186. University of Illinois Press, Urbana (1970)

39. Van Genuchten, M.T.: Calculating the unsaturated hydraulic conductivity with a new closed-form analytical model. Researh Reprot - Water Resour. Program, Dep. of Civ. Eng., Princeton Univ., Princeton, NJ (1978)
40. Von Neumann, J., Burks, A.W., et al.: Theory of self-reproducing automata. IEEE Trans. Neural Networks 5(1), 3–14 (1966)
41. Willebeek-LeMair, M.H., Reeves, A.P.: Strategies for dynamic load balancing on highly parallel computers. IEEE Trans. Parallel Distrib. Syst. 4(9), 979–993 (1993)
42. Wolfram, S.: A New Kind of Science. Wolfram Media Inc., Champaign (2002)

Two Possible AI-Related Paths for Bottom-Up Synthetic Cell Research

Pasquale Stano$^{(\boxtimes)}$ (iD)

University of Salento, Lecce, Italy
pasquale.stano@unisalento.it

Abstract. Advancements in bottom-up synthetic biology (i.e., constructing synthetic cells from scratch) pave the way to new and yet unexplored directions, like: (i) the exploration of semantic aspects of chemical information processing, and (ii) the potential reciprocal inspiration that comes from AI methods and systemic (bio)chemical computation.

Keywords: AI · Chemical intelligence · Embodied AI · Semantic information · Synthetic biology · Synthetic cells

1 The Growing Interest Toward Synthetic Biology as the Wetware Branch of the "Sciences of the Artificial"

The "synthetic method" aims at proving theories of natural phenomena (e.g., life and cognition) by building a proper *artificial system*, which operates autonomously and behaves like a natural system [2]. This methodological procedure is applied in life sciences and cognitive sciences to verify or reject hypotheses about the generative mechanisms underlying complex life-like behavior. Among the possible generative mechanisms that successfully reproduce the targeted phenomenon, purely imitative ones should be discarded – because of their poor relevance for a true scientific understanding, while mechanisms grounded on an organization that mirrors the biosystem organization should be favoured [4].

Artificial systems (artifacts), then, are at the core of the "sciences of the artificial" [17], whereby three branches can be recognized: hardware, software, and wetware, corresponding to robotics, artificial intelligence (AI), and Synthetic Biology (SB). We refer to the SB corner devoted to the investigation of the emergence of life [10,11,18] and cognition [1,6] by the construction of artificial models of the simplest living cells.

The construction of synthetic cells (SCs) becomes, in this respect, a key technology for the sciences of the artificial. Numerous and impressive are the technical developments in the field, triggered by a somehow unexpected recent growth of interest [7]. SCs, it is important to remark, are not yet "alive", but since the most recent and exciting research has endowed SCs with several different and complex functions it is possible to make partially speculative anticipations about

J. J. Schneider et al. (Eds.): WIVACE 2021, CCIS 1722, pp. 97–100, 2022.
https://doi.org/10.1007/978-3-031-23929-8_9

the next directions. Here we would like to sketch some interesting opportunities for future investigations. The starting point comes from the following consideration: the chemical turn in the synthetic modeling of life and cognition (realized by SB approaches) relies on the unparalleled feature of "molecular embodiment", which can progressively blur the distinction between data and processes, information and machine (and, ultimately, object and subject), favouring a circularly and operationally closed merger – typical of living systems and their autopoiesis [14].

2   Possible Directions

Current technical capabilities already foresees non trivial bottom-up SC designs, like the implementation of signal-response mechanisms, bidirectional communication [8], and cell-free genetic circuits and logic gates [16]. It is possible to conceive specific programs such as (i) exploring the semantic aspects of chemical information processing, and (ii) realizing new chemical devices prompted by the reciprocal inspiration between AI methods and systemic (bio)chemical computation.

Communicating SCs and the MacKay theory of semantic information: A Useful Concept?

One of the most attractive SC research line refers to the capacity of SCs of communicating with other SCs, or, intriguingly, with biological cells. The hybrid scenario (synthetic/natural interfacing) resembles, in several aspects, well known situations in the realm of artificial agents, namely the dichotomy between natural intelligence and AI, or between biological bodies and robots. The importance of communicating SCs is not only theoretical: it can represent the operational layer to interface SCs and natural cells, thus contributing to nanomedicine smart vectors[1], and also for developing coordinated behavior in populations of SCs (to generate a next-level of organization).

When discussed from the molecular biology viewpoint, chemical communication (signalling) is described qualitatively or semi-quantitatively according to the usual language of experimental biology. On the other hand, it is interesting to note that from a more rigorous engineering viewpoint, communicating SCs can be described according to the classic Shannon information and communication theory. In particular, a sub-field of communication engineering has recently born, called "molecular communication". It is dedicated to the development of mathematical models of such special type of messaging carried out by molecules. However, there is still another approach that can find in SC research a powerful allied for future and novel investigations. We refer to the MacKay semantic

---

[1] We refer to scenarios whereby properly built SCs act as smart nanomachine for the production of therapeutics via chemical synthesis, inside a biological body. Thanks to their artificially built cognitive traits, SCs could recognize and interact with natural cells via communicative operations. Their entire behavior would result intelligent, due to the capacity of correctly interpreting their surroundings.

theory of information, which is old as the Shannon theory, but was overtook in early days of cybernetics by the Shannon theory because when the meaning of a message is put aside, communication can be modelled by the exact and objective tools of mathematics (while the meaning of messages was labelled as subjective, and context-dependent) [9].

According to MacKay [12], the technical usage of the word 'meaning' lies in the *selective function*, operated by an event that can be detected by an organism (or a machine), upon the ensemble of transition probabilities which characterizes the organism (or machine) behaviour. Notice that the so-defined meaning entails a relation between the message source and the recipient [15]. The semantic theory of information ultimately associates the message meaning to the dynamics it elicits within the receiver, putting on the stage the receiver *internal organization* and its adaptation to the perturbations received from the environment. A new open question emerges: can SB approaches, the ones rooted in self-organization theories (like the bottom-up SCs), be utilized to explore experimentally the MacKay theory, moving beyond the usual approach in term of information as the amount of transferred bit? [13]

Bio-inspired AI and AI-Inspired (Bio)Chemical Intelligence: A Viable Two-Way Path?

New SC designs can emerge from the reciprocal influence between bio-inspired AI strategies and AI-inspired (bio)chemical circuitry. We know that biological structures and functions have profoundly inspired AI. Biologically inspired computing methods have literally revolutionized the field. Approaches such as neural networks, genetic algorithms, and membrane computing – just to mention a few names – are typical examples of bio-inspired computing. No doubts that those approaches are contributing to the impressive success of AI methods we witness in present days. On the other hand, the new experimental horizon provided by advancements in SB can pave the way to novel directions, for example those whereby current AI methods (bio-inspired or not) trigger or guide the construction of artificial (bio)chemical intelligent systems. An example can be the implantation of chemical neural network inside SCs. The latter will be built in the laboratory and can display unique features not available to the silicon AI. We foresee a field which is shaped by inspirational and bidirectional flows between biology and computing.

3   Towards SB-AI Research Programs

It is highly probable that future directions in SC research will be devoted to intertwine and interbreed SB and AI. Here we have sketched two possible paths, which emerge from personal interests in minimal cognition and autopoietic chemical AI [3,5]. Others are also conceivable.

Acknowledgements. I am grateful to Prof. Maurizio Magarini (Politecnico di Milano, Milan, Italy) and Prof. Luisa Damiano (IULM, Milan, Italy) for useful discussions.

References

1. Bitbol, M., Luisi, P.L.: Autopoiesis with or without cognition: defining life at its edge. J. R. Soc. Interface **1**(1), 99–107 (2004). https://doi.org/10.1098/rsif.2004.0012
2. Damiano, L., Hiolle, A., Cañamero, L.: Grounding synthetic knowledge. In: Lenaerts, T., Giacobini, M., Bersini, H., Bourgine, P., Dorigo, M., Doursat, R. (eds.) Advances in Artificial Life, ECAL 2011, pp. 200–207. MIT Press, Cambridge (2011)
3. Damiano, L., Stano, P.: Synthetic biology and artificial intelligence. grounding a cross-disciplinary approach to the synthetic exploration of (embodied) cognition. Complex Syst. **27**, 199–228 (2018). https://doi.org/10.25088/ComplexSystems.27.3.199
4. Damiano, L., Stano, P.: On the "life-likeness" of synthetic cells. Front. Bioeng. Biotechnol. **8**, 953 (2020). https://doi.org/10.3389/fbioe.2020.00953
5. Damiano, L., Stano, P.: A wetware embodied AI? Towards an autopoietic organizational approach grounded in synthetic biology. Front. Bioeng. Biotechnol. **9**, 873 (2021). https://doi.org/10.3389/fbioe.2021.724023
6. Hanczyc, M.M., Ikegami, T.: Chemical basis for minimal cognition. Artif. Life **16**(3), 233–243 (2010). https://doi.org/10.1162/artl_a_00002
7. Ivanov, I., et al.: Bottom-up synthesis of artificial cells: recent highlights and future challenges. Annu. Rev. Chem. Biomol. Eng. **12**(1), 287–308 (2021). https://doi.org/10.1146/annurev-chembioeng-092220-085918
8. Lentini, R., et al.: Two-way chemical communication between artificial and natural cells. ACS Cent. Sci. **3**(2), 117–123 (2017). https://doi.org/10.1021/acscentsci.6b00330
9. Logan, R.K.: What is information?: Why is it relativistic and what is its relationship to materiality, meaning and organization. Information **3**(1), 68–91 (2012). https://doi.org/10.3390/info3010068
10. Luisi, P.L.: Toward the engineering of minimal living cells. Anat. Rec. **268**(3), 208–214 (2002). https://doi.org/10.1002/ar.10155
11. Luisi, P.L., Varela, F.J.: Self-replicating micelles - a chemical version of a minimal autopoietic system. Origins Life Evol. Biosphere **19**(6), 633–643 (1989). https://doi.org/10.1007/BF01808123
12. MacKay, D.M.: Information, Meaning and Organization. MIT Press, Cambridge (1969)
13. Magarini, M., Stano, P.: Synthetic cells engaged in molecular communication: an opportunity for modelling shannon- and semantic-information in the chemical domain. Front. Commun. Netw. **2**, 48 (2021). https://doi.org/10.3389/frcmn.2021.724597
14. Maturana, H.R., Varela, F.J.: Autopoiesis and Cognition: The Realization of the Living. D. Reidel Publishing Company (1980)
15. Pask, G.: An Approach to Cybernetics. Hutchinson & Co., Ltd., London (1961)
16. Pieters, P.A., et al.: Cell-free characterization of coherent feed-forward loop-based synthetic genetic circuits. ACS Synth. Biol. **10**(6), 1406–1416 (2021). https://doi.org/10.1021/acssynbio.1c00024
17. Simon, H.A.: The Sciences of the Artificial. MIT Press, Cambridge (1996)
18. Szostak, J.W., Bartel, D.P., Luisi, P.L.: Synthesizing life. Nature **409**(6818), 387–390 (2001). https://doi.org/10.1038/35053176

Robot Systems

A Hybrid Control System Architecture for a Mobile Robot to Provide an Energy-Efficient and Fast Data Processing

Uladzislau Sychou[1]([✉]) [iD], Ryhor Prakapovich[1] [iD], Sergej Kapustjan[2] [iD], and Vladimir Shabanov[3]

[1] Laboratory of Robotic Systems, United Institute of Informatics Problems of National Academy of Sciences of Belarus, Minsk, Belarus
vsychyov@robotics.by
[2] Southern Scientific Centre of the Russian Academy of Sciences, SSC RAS, Rostov On Don, Russia
[3] Southern Federal University, Taganrog, Russia

Abstract. The research is aimed to improve a controller for spherical mobile robots. The main issues of the existing controller are related to the computational capabilities of a low-power digital controller and the slow handling of sensor data. A survey of suitable contemporary technologies and approaches to spherical robot control was done. A hybrid analogue-digital architecture is considered the best way to control the robot. It is concluded, that exactly hybrid controller, which combines specialised analogue accelerators with multi-purpose digital controllers can be the most power-efficient, compact and fast architecture for bio-inspired robots.

Keywords: Mobile robot · Spherical robot · Analogue controller · Reservoir computing · Chaotic oscillator

1 Introduction

Since the first robots appeared the attempts to use them wider have been continuing. At present, robots are used in industry, agriculture, construction, mining, and in many other spheres. Difficulties in the development of robotics are overcome using compact controllers with high computational efficiency and better sensors. Moreover, batteries, motor drivers and power electronics become more efficient. Modern hardware enables different technologies of programming, including artificial neural networks (ANN).

Nevertheless, the most common robots remain classical ones, with conventional actuators, centralized controllers and dimensions from dozens of centimetres to meters. Often, such robots are not fitted to work in human-environment or tough conditions due to specialised actuators and algorithms are required. In that case, bio-inspired robotics can be considered as the prospective way of robotics evolution. One of the robots, which is based on not commonly used technologies, is the spherical robot developed by authors.

J. J. Schneider et al. (Eds.): WIVACE 2021, CCIS 1722, pp. 103–108, 2022.
https://doi.org/10.1007/978-3-031-23929-8_10

2   Spherical Mobile Robot

The spherical mobile robot (see Fig. 1) is twenty-five centimetres plastic ball with a core module inside. The hermetic enclosure enables the robot to work in an inimical environment. The spherical form allows moving on the areas that are not suitable for wheeled or tracked robots [1]. There are many kinds of spherical robots. The most common way to drive it is based on an inverted pendulum inside the sphere. In such a case, the centre of mass of the robot is not in the geometrical centre of the sphere. The robot moves only when the pendulum inside changes its position. Consequently, the robot can't move without energising motors constantly and even short discontinuing acceleration can't make it roll freely like a ball when motors are turned off. Another common type of spherical robot employs a wheeled or tracked cart inside. The rotation of the cartwheels transmits on the enclosure. Such a robot commonly remains the same disadvantages.

The proposed robot has a module with a controller, battery, and six servos. The module is placed in the enclosure and connected with it by the six links of a parallel manipulator. The module can move in any direction. Biasing the module relative to the centre of the sphere yields torque. Oppositely, placing the module in the centre inside the sphere makes the robot balanced enough to allow it to move freely. As a result, on a plane surface, the robot can roll even after short torque. The low resistance to rolling is considered as the main feature of the robot, which can help save energy.

Fig. 1. The general form of the spherical mobile robot: the prototype (at left) and the model in Simulink (at right)

To drive the robot the controller moves links synchronously and uses sensors feedback. The design of the robot makes remote control by the operator difficult due to a camera cannot be mounted inside and provide enough information in the common case. As result, the optimal strategy of using the spherical robot to solve real-world tasks is autonomous operation.

While tuning the controller, several difficulties were noticed. In particular, a low-power microcontroller is not enough efficient to perform real-time mathematics and trigonometric calculations. The usage of standard sensors with digital interfaces brings signal propagation delays. The absence of operator control implies algorithms for autonomous exploration, patch planning, machine learning, etc. Limited dimensions hindered heat dissipation restrain the usage of common architectures. The technologies that can be useful for spherical robot control are considered below.

3   Analogue Electronic Computers

Analogue electronic controllers are used since the first half of the twenty century in many spheres where electronic modules must be compact, energy-efficient and fast [2]. At present, analogue computers can provide remarkable computational capability in comparison with digital computers, especially low-power microcontrollers [3]. The number of the classical techniques that can be implemented on an analogue base includes PID-controller [4], LQR-controller [5], Kalman filter [6]. The principles of bionics and analogue electronics are the background of BEAM robotics [7, 8] – the approach that implies the usage of simple electronic circuits to control the movements of a robot.

Besides the common control tasks, analogue electronics can solve more complicated tasks. For instance, [9] describes a machine-learning system on chip (SoC). A deep-learning engine with computational power is about 1 TOPS/W was reported in [10]. Moreover, this engine was the basis of IC that was produced in 0.13 um technology. Breakthrough in analogue electronics is memristor [11]. It can improve the flexibility of analogue circuits to implement more efficient ANNs, controllers, filters [6].

To make a robot more autonomous it is not necessary to use complicated algorithms. One of the variants of autonomous behaviour is exploration locomotion, which is intrinsic to many living things. And exploration algorithms were historically the first that can be described as bio-inspired. For instance, the Braitenberg vehicle, being extremely simple, shows complicated behaviour, which even can appear intelligent [12].

In this context should be mentioned the fact that spherical form is absent amongst common shapes of animal bodies. Presumably, the main reason is that the described principle of motion is good enough only for ideal spherical shapes with well-balanced mass. Such properties are hard to be achieved for living creatures. However, hazelnut or walnut, as well as apple and many other fruits and nuts can roll back from the three to grow in free space due to the spherical form. Consequently, the advantages of the spherical form are employed in nature.

4   Chaotic Oscillators and Reservoir Computing Framework

The prospective way to perform calculations is by employing the inherent properties of complex systems. In particular, there are systems, which can produce chaotic oscillations. These systems have such important features as high dimensionality, nonlinearity, fading memory [13].

Due to these features, chaotic oscillators are applied in robotics to the synthesis of motion trajectories for robot exploration [14]. In machine learning, Chua chaotic oscillator was the basis of an ANN that was successfully applied for pattern recognition of handwritten digits [15] and to construct an associative memory architecture [16].

More widely, using the inherent properties of systems is the main idea of reservoir computing (RC) [17]. RC is a machine learning technique that applies nonlinear dynamical systems as a computational core. Some RC techniques involve chaotic oscillators [18]. An important feature of reservoirs is the ability to be built in as a hardware accelerator to solve some control and machine learning tasks.

5   The Dynamics of Spherical Robot

The considered robot was developed as a Simulink model and then embodied as a working prototype. Observations confirm the adequacy of the model. The robot is controlled by the position of the core module that depends on longitude, latitude, and radius (bias from the geometrical centre). Let us consider how the robot reacts to control signals without feedback. The core module was biased fourteen millimetres from the centre. The latitude was fixed at zero level and the longitude has been changing constantly yield circle movement of the core module inside the sphere. However, hasn't been controlled relating to sensors, the robot moves not straight (Fig. 2).

Angular velocities are plotted in Fig. 3. The graph shows that initial movement along the y-axis is followed by a little displacement along the z-axis. The reason is the manner how the parallel manipulator works. All shafts must rotate synchronously as shown in Fig. 2, right graph. Consequently, it is not always possible to avoid appearing torque in an undesirable direction. The displacement of the core along the z and y directions produces torque along the x-axis, which is changing the direction of rolling and robot orientation.

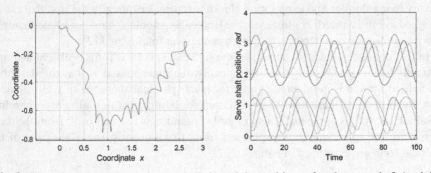

Fig. 2. The robot movement trajectory (at left) and the positions of each motor shaft (at right)

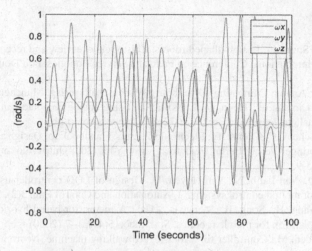

Fig. 3. Angular velocities along three axes.

6 The Hybrid Architecture of the Control System

The acceleration is considered as the input of the control system. The acceleration can be produced by moving the core as well as by the external impact. The desired position of the core in spherical coordinates is the output of the controller. Feedback loops involve accelerometers and gyroscopes. The control system consists of two main parts. The first one, "linear part", is based on linear equations that describe the movement of the body depending on accelerations. This part is based on filters, integrators, regulators, which can be tuned using classical control theory. The second part of the system is needed to determine the core position and trajectory using the desirable acceleration of the robot and vice versa taking into account the particular qualities of the parallel manipulator. Reservoir computing is considered as the prospective technology to build this part of the controller.

To summarise, the most promising architecture is that combines specialized analogues modules with a digital unit. Separate modules can serve as filters, controllers, reservoir computers, path planners. The approach is close to the one described in [6]. In the cited research the PID-controller and Kalman filter were implemented as separate analogue accelerators that can be tuned using memristors. It is allowed to achieve outstanding power efficiency and reaction speed. At the same time, it was claimed that only the digital part of the controller should run intelligent functions. However, the considered examples allow us to conclude that even complicated machine-learning and navigation tasks can be accomplished by analogue accelerators.

Acknowledgement. This work has been supported by the grant BRFFI-RFFI No. F18R-229.

References

1. Ylikorpi, T, Suomela, J.: Ball-shaped robots: a historical overview and recent developments at TKK. In: International Conference on Field and Service Robotics. Port Douglas, Australia, p. 6 (2005)
2. Ulmann, B.: Analog Computing. Oldenbourg Wissenschaftsverlag, München (2013)
3. Köppel, S., Ulmann, B., Heimann, L., Killat, D.:Using analog computers in today's largest computational challenges. arXiv:2102.07268v2 [physics.comp-ph]
4. Tietze, U., Schenk, C., Gamm, E.: Electronic Circuits: Handbook for Design and Application. 2nd edn. Springer-Verlag Berlin Heidelberg, New York (2008).https://doi.org/10.1007/978-3-540-78655-9
5. Cuong, N.D., Van Lanh, N., Van Huyen, D.: Design of LQG controller using operational amplifiers for motion control systems. J. Automation and Control Eng. 3(2), 157-163 (2015)
6. Chen, B., Yang, H., Song, B., Meng, D., et al.: A memristor-based hybrid analog-digital computing platform for mobile robotics. Sci. Robot. 5(47), 1–7 (2020)
7. Still, S., Tilden, M.: Controller for a four-legged walking machine. Neuromorphic Systems 138–148 (1998)
8. Hasslacher, B., Tilden, M.W.: Living machines. Robotics Auton. Syst. 15(1-2), 143-169 (1995)
9. Hasler, J.: Analog abstraction, computation, and numerical analysis. In: 2018 IEEE International Symposium on Circuits and Systems (ISCAS), Florence, pp. 1–5 (2018)
10. Lu, J., Young, S., Arel, I., Holleman, J.: A 1 TOPS/W analog deep machine-learning engine with floating-gate storage in 0.13 um CMOS. IEEE J. Solid-State Circuits 50, 270–281 (2015)
11. Chua, L.O., Sirakoulis, G.C., Adamatzky, A.: Handbook of Memristor Networks. Springer (2019). https://doi.org/10.1007/978-3-319-76375-0
12. Braitenberg, V.: Vehicles, Experiments in Synthetic Psychology. MIT Press, Cambridge, Mass (1984)
13. Moon, F.: Chaotic Vibrations. John Wiley&Son (2004)
14. Sooraska, P., Klomkarn, P.: "No-CPU" chaotic robots: from classroom to commerce. Circuits and Systems Magazine, IEEE 10, 46–53 (2010)
15. Baird, B., Hirsch, M.W., Eeckman, F.: A neural network associative memory for handwritten character recognition using multiple Chua characters. IEEE Trans. Circuits Syst. II: Analog and Digital Signal Processing 40, 667–674 (1993)
16. Jankowski, S., Londei, A., Mazur, C., Lozowski, A.: Synchronization and association in a large network of coupled Chua's circuits. Int. J. of Electronics 79, 823–828 (1995)
17. Nakajima, K., Fischer, I.: Reservoir computing. Theory, Physical Implementations, and Applications. Springer (2021). https://doi.org/10.1007/978-981-13-1687-6
18. Tanaka, G., Yamane, T., Heroux, J.B., Nakane, R., et al.: Recent advances in physical reservoir computing: a review. Neural Netw. 115, 100–123 (2019)

On the Evolution of Mechanisms
for Collective Decision Making
in a Swarm of Robots

Ahmed Almansoori[1,2]([⊠]) [iD], Muhanad Alkilabi[1,2] [iD], Jean-Noël Colin[1] [iD],
and Elio Tuci[1] [iD]

[1] University of Namur, Namur, Belgium
{ahamed.almansoori,jean-noel.colin,elio.tuci}@unamur.be
[2] University of Kerbala, Karbala, Iraq
{Ahmed.kamil,muhanad.hayder}@uokerbala.edu.iq
https://www.unamur.be/info/

Abstract. Collective decision-making refers to a process of generating a
group decision which cannot be attributed to any agent in the group. In
swarm robotics, the individual mechanisms for collective decision making
are generally hand-designed and limited to a restricted set of solutions
based on the voter or the majority model. This study demonstrates that
it is possible to take an alternative approach in which the individual
mechanisms are implemented using artificial neural network controllers
automatically synthesised using evolutionary computation techniques.
We qualitatively describe the group dynamics underpinning the collective process leading to consensus. Moreover, this study demonstrates
the evolutionary-tailored mechanisms do not follow the principles of the
classic hand-coded solutions.

Keywords: Swarm robotics · Collective decision making · Automatic
design

1 Introduction

Swarm robotics is a particular type of multi-robot systems in which each robot
has its own controller, perception is local and communication is based on spatial
proximity [6]. The swarm's designer operates at the individual level, by providing each robot the mechanisms that generate its behaviour. The group-level
or swarm response emerges, through a self-organisation process, from the interactions between the robots and their social and physical environment. Due to
the distributed and random nature of this self-organisation process, it is noto-
riously difficult for the designer to predict which set of individual actions leads
to the emergence of the desired collective response [2]. In this study, we focus

A. Almansoori—Thanks UNamur for the financial support. M. Alkilabi—Thanks the
research institute naXys for the financial support.

J. J. Schneider et al. (Eds.): WIVACE 2021, CCIS 1722, pp. 109–120, 2022.
https://doi.org/10.1007/978-3-031-23929-8_11

on the design of individual mechanisms for collective decision making: that is, a decision process in which none of the individuals can be accounted for the group choice once a collective decision is made [17]. In nature, similar processes can be observed in social insects, which collectively choose a foraging or a nesting site without any agent knowing the quality of all available options [3]. The swarm robotics literature has shown that in multiple contexts, a consensus can be reached by swarms in which individuals change opinion by following the rules of the voter model, whereby agents change opinion copying a random neighbour, or of the majority model, whereby agents change opinion to the option held by the majority of a group of neighbours [4,5,13,18,19]. In this study, we take a complementary approach. Instead of making strong assumptions (e.g., voter or majority model) on the nature of the individual mechanisms underpinning collection decision making processes and showing their effectiveness in specific swarm robotics tasks, we select an already investigated scenario and we exploit the evolutionary robotics methodology to synthesise individual neural mechanisms for options selection. The evolutionary approach is based on the use of evolutionary computation techniques to synthesise artificial neural networks as robots' controller [15]. We chose a scenario in which the robots explored a close arena with the floor made of black and white tiles. The swarm has to reach a consensus on which type (black or white) of tiles covers the majority of the arena floor. This scenario has been originally investigated in [16] where the authors tested the effectiveness of three different mechanisms for options selection (a weighted voter model, the majority model, direct comparison) both with simulated and physical robots. In [14], the authors replicated the study described in [16] to test the effectiveness of a blockchain-based smart contract to protect the collective decision making process from agents (i.e., byzantine robots) acting in order to disrupt the decision process. The blockchain-based layer operates on top of the decision making mechanisms based on the majority rules. In [8], the collective perception scenario described in [16] has been transformed into a multi-feature collective decision making problem, in which the simulated and physical robots reach consensus using hand-designed options selection mechanisms.

To the best of our knowledge, this is the first study demonstrating that the evolutionary approach can be successfully used to synthesise the robots mechanisms in the above described collective perception task in which the group consensus has to be reached with partial individual knowledge and communication limited to the closest neighbour. We provide a preliminary analysis of the evolved collective behaviour showing that if we automate the design of individual mechanisms with a process that evaluates the group performance without imposing specific solutions at the operational level, the evolutionary-tailored mechanisms do not resemble what commonly assumed necessary to reach consensus. In our case, we show that the evolved individual mechanisms for options selection do not follow the rule of the voter model. This result demonstrates that the evolutionary approach generates alternative solutions that may improve the effectiveness of the collective decision making process through individual mechanisms that are task-specific instead of being of task-independent. Moreover, the plasticity, resiliency, and generalisation capabilities of artificial neural networks, may contribute to improve the robustness and the adaptivety of the swarm with respect to a larger spectrum of sources of environmental variation.

As mentioned above, this is one of the first study using the evolutionary approach to the design of options selection mechanisms for robots of a swarm engaged in a collective decision making scenario. The authors in [12] explored a similar scenario with the evolutionary methods, but with the support of global communication which significantly reduces the complexity of the decision process and its significance for the swarm robotics community. In our scenario, each robot can communicate its opinion only locally to the nearest neighbour and interactions are possible only between agents within 50 cm to each other. To reduce the complexity of the evolutionary task, we take a modular approach in which the movement of the robots is determined by an hand-coded algorithm for pseudo-random walk, while the neural mechanisms for options selection are synthesised using artificial evolution. The modularisation of the control structure does not limit in any way the significance of our results for the swarm robotics community. A detailed analysis of the results of this research work is provided in Sect. 3, while a discussion on the contributions of this study and on the possible developments for future work are illustrated in Sect. 4.

2 Methods

(a)   (b)

Fig. 1. (a) The simulated collective decision-making scenario. (b) Image of a physical e-puck robot.

We study a collective decision-making scenario that has been described for the first time in [16]. In this scenario, a swarm of 20 robots navigate a close arena of 2×2 m. The arena floor is made of black and white tiles, 10×10 cm each, which are randomly distributed on the arena floor (see Fig. 1a). During the design phase, the swarm experiences two types of environment: a) the black-dominant environment in which 55% of the tiles are black and 45% are white; and b) the white-dominant environment in which 55% of the tiles are white and 45% are black. As in [16], the task of the swarm is to reach a consensus on the type of environment on which they are placed. In this study, consensus refers to a state in which all the robots of the swarm share the same opinion on which tiles' colour covers the largest portion of the arena floor. We have studied two experimental conditions: the comm. experimental condition, which refers to an experimental setup in which the robots can exchange their opinion through communication. To show that consensus can not be reached unless the robots exploit the collective intelligence, we have replicated the study in the no-comm. conditions in which the robots can not communicate their opinion to each other.

During evaluation, the robots move according to a isotropic random walk, with a fixed step length (5 s., at 20 cm/s), and turning angles chosen from a wrapped Cauchy probability distribution characterised by the following PDF:

$$f(\theta, \mu, \rho) = \frac{1}{2\pi} \frac{1 - \rho^2}{1 + \rho^2 - 2\rho \cos(\theta - \mu)}, \ 0 < \rho < 1, \tag{1}$$

where $\mu = 0$ is the average value of the distribution, and ρ determines the distribution skewness (see [9]). For $\rho = 0$ the distribution becomes uniform and provides no correlation between consecutive movements, while for $\rho = 1$ a Dirac distribution is obtained, corresponding to straight-line motion. In this study $\rho = 0.5$. While moving around, the robots continuously perform an obstacle avoidance behaviour. To perform obstacle avoidance, first a robot detects an obstacle, then stops and keeps on changing its headings of a randomly chosen angle uniformly drawn in $[0, \pi]$ until no obstacles are perceived. Given the robots' pseudo-random walk and the random distribution of black and white tiles on the arena floor, we estimated, by simulating multiple times the task, that each robot explores on average only about 18% of the arena tiles during each evaluation period (i.e., 200 s.). Thus, robots have to develop their opinion based on partial knowledge of the environment and by exploiting the collective intelligence through communication.

The swarm controllers are evolved in a simulation environment which models some of the hardware characteristics of the e-puck2 robots (see Fig. 1b), small wheeled cylindrical robots, 70 mm diameter, equipped with a variety of sensors, and whose mobility is ensured by a differential drive system (see [11] for details). In this study, the simulated e-pucks are equipped with infra-red sensors, placed all around the robot's body, the floor sensor, placed underneath the robot chassis. The signal of the infrared sensors is a function of the distance between the robot and any perceived obstacle (in this task, an obstacle can be another robot or the arena walls). The floor sensors return 0 when the centre of the robot base is on a black tile, and 1 when is on a white tile. For the communication, we simply assume that whenever two robots are at less than 50 cm from each other, a 1 bit signal sent by one agent can be perceived by the other. Each robot signals its opinion for the entire duration of the evaluation. The signal sender communicates its current opinion which either black-dominant (the signal sent is 0) or white-dominant (the signal sent is 1). This type of communication can be reliably implemented on physical e-puck2 with the range&bearing board. Concerning the function that updates the position of the robots within the environment,

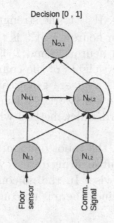

Fig. 2. Architecture of the neural network that control the agents

we employed the Differential Drive Kinematics equations, as presented in [7]. To compensate for the simulation-reality gap, 10% uniform noise is added to all sensor readings, the motor outputs and the position of the robot [10].

2.1 The Robot Controller

In the comm. condition, each robot is controlled by a continuous time recurrent neural network (CTRNN) [1]. The neural network has a multi-layer topology, as shown in Fig. 2: neurons $N_{I,1}$ and $N_{I,2}$ take input from the robot's floor sensor and the eventual communication signal (1 for white-dominant, 0 for black-dominant, and 0.5 whenever there is no other robots at less than 50 cm from the receiver), neuron $N_{O,1}$ is used to set the robot opinion, and neurons $N_{H,1}$ and $N_{H,2}$ form a fully recurrent continuous time hidden layer. The input neurons are simple relay units, while the output neuron is governed by the following equations:

$$o = \sigma(O + \beta^O), \tag{2}$$

$$O = \sum_{i=1}^{2} W_i^O \, \sigma(H_i + \beta_i^H), \tag{3}$$

$$\sigma(z) = (1 + e^{-z})^{-1}, \tag{4}$$

where, using terms derived from an analogy with real neurons, O and H_i are the cell potentials of respectively output neuron and hidden neuron i, β^O and β^H are bias terms, W_i^O is the strength of the synaptic connection from hidden neuron i to output neuron, and o_j and $h_i = \sigma\,(H_i + \beta_i)$ are the firing rates. The hidden units are governed by the following equation:

$$\tau_j \dot{H}_j = -H_j + \sum_{i=1}^{2} W_{ij}^H \sigma(H_i + \beta_i^H) + \sum_{i=1}^{2} W_{ij}^I I_i, \tag{5}$$

where τ_j is the decay constant, W_{ij}^H is the strength of the synaptic connection from hidden neuron i to hidden neuron j, W_{ij}^I is the strength of the connection from input neuron i to hidden neuron j, and I_i is the intensity of the sensory perturbation on neuron i. The weights of the connection between neurons, the bias terms and the decay constants are genetically encoded parameters. Cell potentials are set to 0 each time a network is initialised or reset. State equations are integrated using the forward Euler method with an integration step-size of 0.1 s.

Neuron $N_{O,1}$ is used to set the robot opinion, which corresponds to 1 (i.e., white-dominant) when the neuron firing rate is above the threshold 0.5, and 0 (i.e., black-dominant) otherwise. In the no-comm. condition, $N_{I,2}$ and the connections $\omega_{2,j}^I$ are removed since there is no communication signal.

2.2   The Evolutionary Algorithm and the Fitness Function

A simple evolutionary algorithm using linear ranking is employed to set the parameters of the networks. The population contains 100 genotypes. Generations following the first one are produced by a combination of selection with elitism, and mutation. For each new generation, the highest scoring individuals ("the elite") from the previous generation is retained unchanged. The remainder of the new population is generated by binary tournament selection from the 70 best individuals of the old population. In the comm. condition, each genotype is a vector comprising 15 real values (10 connections, 2 decay constants, 3 bias terms). In the no-comm. condition, each genotype is a vector comprising 13 real values (8 connections, 2 decay constants, 3 bias terms). Initially, a random population of vectors is generated by initialising each component of each genotype to values chosen uniformly random from the range $[0,1]$. New genotypes, except "the elite", are produced by applying mutation, which entails a random Gaussian offset applied to each real-valued vector component encoded in the genotype, with a probability of 0.03. The mean of the Gaussian is 0, and its standard deviation is 0.1. During evolution, all vector component values are constrained to remain within the range $[0,1]$.

At the beginning of each evaluation trial, each genotype is decoded into a neuro-controller. Then, the controller is cloned on each of the $R = 20$ robots forming the swarm (i.e., we use homogeneous swarms). The robots are randomly placed in the arena with a randomly chosen orientation. Each trial differs from the others in the initialisation of the random number generator, which influences the robots' initial position and orientation, and the noise added to motors and sensors. Within a trial, the swarm life-span is 200 s ($T = 2000$ simulation cycles). The fitness of a genotype is its average swarm evaluation score after it has been assessed 2 times, in each type of environment (i.e., the black-dominant, and the white-dominant environment) for a total of 4 trials. In each trial e, the opinion of robot r is evaluated at every time t (i.e., O_t^r) for every robot in the swarm. Finally, the swarm is rewarded by an evaluation function F_e which is computed as follows:

$$F_e = \begin{cases} \frac{T}{2} \sum_{t=T/2}^{T} \sum_{r=1}^{R} o_t^r \text{ if swarm located in a white-dominant env.} \\ \frac{T}{2} \sum_{t=T/2}^{T} \sum_{r=1}^{R} (1 - o_t^r) \text{ if swarm located in a black-dominant env.} \end{cases} \quad (6)$$

Note that, within each trial, the fitness score is computed from the simulation cycle T = 1000 to the end of the trial. This is to exclude from the fitness score the effects of the inevitable fluctuations in the agents' opinion that are observed at the beginning of each trial when the agents controller is in the "default" initial state.

3   Results

We have performed 20 differently seeded evolutionary simulation runs for each experimental condition. We remind the reader that we have two experimental conditions, the comm. condition where each robot can communicate with the closest robot if its distance is less than 50 cm; and the no-comm. condition where there is no communication between the robots. Each run of each condition lasts 500 generations. In each run, each genotype is evaluated for 4 trials. A trial starts when the robots are randomly placed within the arena and lasts for 200 s. (i.e., 2000 simulation time-steps). In half of the trials, the robots experience a black-dominant, and in the other half the white-dominant environment. Each trail is an independent event, since at its beginning the robots' controller is reset (see Sect. 2.1 for details). At the end of the evolutionary phase, for each condition, we have re-evaluated the best genotype of each of the last 100 generations of each evolutionary run. In this re-evaluation phase, the performance of each best genotype is estimated on 100 trials (i.e., 50 in a black-dominant, and 50 in a white-dominant environment), using Eq. 6 as evaluation metrics.

In the no-comm. condition, the performances of these group never rose above 80%. This indicates that none of these groups managed to successfully solve this collective decision-making task in totally. In the comm. condition, several genotypes of different evolutionary runs performed very well, with a fitness score very close to the maximum value. This indicates that every run managed to generate groups that can successfully accomplish this task. In the remaining of this section, we show the results of these and of further post-evaluation tests on the very best genotype of each condition. These tests are meant to illustrate some of the characteristics of the best group performance. Although due to shortage of space, we discuss only a single group performance, several successful groups underwent these post-evaluation tests with very similar results. Thus, the results discussed below are representative of the behaviour of several best-evolved groups in each condition.

During the 100 trials of the first post-evaluation test, we observed that all the robots of both best groups of the comm. and of the no-comm. conditions have a genetic predisposition at the beginning of each trial to chose the black-dominant or the white-dominant opinion.

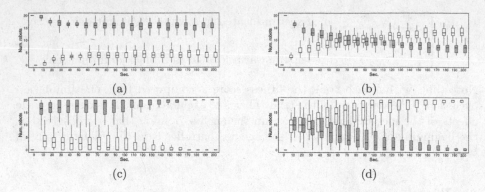

Fig. 3. Box plots showing the number of robots with the opinion white-dominant (white boxes), and those with the black-dominant (grey boxes) every 10 s, during 100 trials in both environments (50 in white-dominant and 50 in black-dominant), for the best group of: (a,b) the no-comm. condition; and (c,d) the comm. condition. Each point in the box refers to the group performance in a single trial. Boxes represent the interquartile range of the data, while horizontal bars inside the boxes mark the median value. The whiskers extend to the most extreme data points within 1.5 times the interquartile range from the box. (Color figure online)

The dynamic of both best groups of the comm. and of the no-comm. conditions are illustrated in Fig. 3 which shows the number of robots with the white-dominant (see white boxes in Fig. 3) and with the black-dominant (see grey boxes in Fig. 3) opinion every 10 s, during 100 trials (50 in white-dominant and 50 in black-dominant), for the best group of the no-comm. condition (Fig. 3a, and 3b) and the comm. condition (Fig. 3c, and 3d). In the no-comm. condition, robots have a genetic bias for the black-dominant opinion. Thus, at trial start in the black-dominant environment, they all share the correct opinion (see Fig. 3a at 0 s). However, few time-steps after start, some of the robots, induced by the perceptual experience of white tiles, they change opinion to white-dominant. The proportion of robots with a white-dominant opinion remains relatively stable until the end of the trial, thus stopping the swarm to achieve a consensus on the black-dominant opinion (see Fig. 3a from 10 s, to 200 s).

When the robots experience a white-dominant environment, the effect of the environment generates different dynamics compared to those observed in the black-dominant environment. As shown in Fig. 3b, the robots progressively change to the white-dominant opinion until about 80 s when both opinions are almost equally represented in the swarm. After that, the group of robots with the white-dominant opinion grows larger than the group of robots with the black-dominant opinion. However, this process never leads to a consensus for the white-dominant opinion. (Figure 3b).

In both environments, the robots keep on changing opinion during the trial, but since there are frequent opinion changes in both ways (i.e., from white-dominant to black-dominant and vice-versa) the median remains rather away

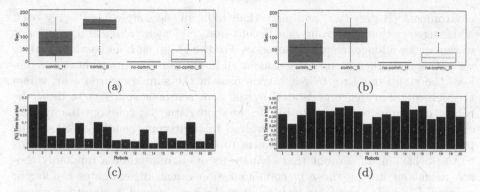

Fig. 4. In (a) and (b), box plots showing the length of the longest periods on consensus over (a) 50 trial in the black-dominant environment, and (b) 50 trials in the white-dominant environment. Grey boxes refer to the best evolved group in the comm. condition, while white boxes refer to the best evolved group in the no-comm. condition. The boxes indicated as S refer to simple environments (i.e., 66% of dominant color and 34% of the another color); those indicated as H refer to the hard environment (i.e., 55% of dominant color and 45% of their another color). In (c) and (d), graphs showing for each robot of the best evolved group of the comm. condition, during a trial in a white-dominant environment, the number of times the floor sensor in (c) and the communication sensor in (d) returns a value which contradicts the current robot opinion, during periods in which the robot's opinion does not change. In both graphs, the numbers inside the bars indicate the number of times the corresponding robot changes opinion within the trial. (Color figure online)

from consensus. Based on this evidence, we claim that without communication this collective decision-making task cannot be accomplished.

In the comm. condition, the group dynamics are quite different, since after the start, the group slowly converge to the correct opinion, in spite of the minimal fluctuations from consensus determined by one or two robots that sporadically and only for a short period of time change opinion (see Fig. 3c, and 3d). In Fig. 4, we will further inspect the behaviour of these successful groups. This will help us to get a better idea of the groups behaviour in this type of environment.

Figure 4a, 4b tells us more about the consensus periods. In particular, Fig. 4a shows the longest periods on consensus over 50 trials in black-dominant environments and Fig. 4b shows the longest periods on consensus over 50 trials in white-dominant environments. The grey boxes refer to the best evolved group in the comm. condition, while white boxes refer to the best evolved group in the no-comm. condition. Each group underwent two sets of 50 trails, the first set in a hard environment (i.e., 55% of white tiles and 45% of black tiles), and the second set in a simple environment (i.e., 66% of white tiles and 34% of black tiles). In both Fig. 4a, and 4b, the results referring to hard environment are indicated with H, while those in the simple environment are indicated with S. For the group with communication, we notice that, for about half of the trails, the length of the longest period on consensus is more than 60 s. in the hard environment (H grey bar) and more than 130 s. in the simple one (S grey bar). This suggests that the group can benefit from easier environmental conditions to remain for a longer time on consensus. For the group without communication, we notice that, there is no consensus at all in the hard environment. Nevertheless, the group manages to reach consensus in the simple environment, where the perceptual evidence generated by the floor sensor is sufficient to drive the robot to a common opinion. In Fig. 4c, we show data that rules out the hypothesis that each robot of the group evolved in the comm. condition follows the principle of the voter model. In swarm robotics, the voter model corresponds to the behaviour of a robot that changes its opinion anytime a randomly chosen neighbour among those in communication range, disseminates a different opinion. Figure 4c shows, for each robot of the best evolved group of the comm. condition, during a trial in a white-dominant environment, the number of times the communication sensor, returns a value which contradicts the current robot opinion, during periods in which the robot's opinion does not change. Since this happens multiple time for each robots, we exclude that the robots are following the rules of the voter model. Data in Fig. 4d let us exclude also the hypothesis that the robots are simply choosing their opinion based on the perception of the floor colour. Figure 4d shows for each robot of the best evolved group of the comm. condition, during a trial in a white-dominant environment, the number of times the floor sensor returns a value which contradicts the current robot opinion, during periods in which the robot's opinion does not change. Again, since this happens multiple times for each robots, we can exclude that the robot opinion is simply determined by the colour of the floor. The robots opinion seems to be generated by mechanisms which follow principles slightly more complex than models in which the robots simply react to either the perception of the floor or to the communication signals. The nature of these mechanisms, which successfully drive the group to a correct consensus in both the black-dominant and the white-dominant environment will be investigate in a future study.

4   Conclusions

The first contribution of this study is to provide a proof-of-concept demonstration of the possibility to synthesise, using evolutionary computation techniques, neuro-controllers for a swarm of robots engaged in a collective decision making tasks requiring consensus of opinions among the robots. We showed that a relatively small dynamic neural network, made of only 5 neurons, is sufficient to generate the mechanisms required by the robots of a homogeneous swarm to collectively decide which colour (between two) is predominant on the arena floor, despite the fact that each robot can only explore a small portion of the floor, and communication is limited both in terms of numbers of different signals received at one time (just one) and in terms of distance (i.e., 50 cm) between sender and receiver. In other words, the neuro-controller supports the development of a collective process which overcomes the individual limitations. This evidence paves the way to an alternative research line which challenges some of the preconceptions of large part of the previous state-of-the-art in swarm robotics by which mechanisms for collective decision making are generally implemented with the voter and/or the majority model. The evolutionary swarm robotics approach, used in this study makes possible to exempt the design from making *a priori* assumptions on the nature of the individual mechanisms required for reaching a consensus in collective decision-making tasks. Indeed, our analysis demonstrated that, in the comm. condition, none of the robots changes its opinion based on the rules of the voter model. We have kept the controllers as small as possible to facilitate the analysis of the operational principles that underpins the individual opinion-making process in this collective task. This will be the subject of future work. Moreover, the evolutionary swarm robotics approach offers the possibility to synthesise mechanisms that are evolutionary tailored to the task requirements, with the possibility to improve the effectiveness and adaptability of the swarm behaviour. This work can be extended in multiple different ways. First, we intend to integrate into the neuro-controller both the mechanisms for the individual opinion-making process and those to generate the robots movement, thus dropping the assumption that robots has to make a specific pseudo-random walk during the exploration of the arena. This will allow us to investigate which individual movements support the collective decision-making process. Second, we would like to study scenarios with more than two options, to study how the task scenario influences the individual behaviour.

References

1. Beer, R.D.: A dynamical systems perspective on agent-environment interaction. Art. Intell. **72**, 173–215 (1995)
2. Brambilla, M., Ferrante, E., Birattari, M., Dorigo, M.: Swarm robotics: a review from the swarm engineering perspective. Swarm Intell. **7**(1), 1–41 (2013)
3. Britton, N., Franks, N., Pratt, S., Seeley, T.: Deciding on a new home: how do honeybees agree? Proc. Biol. Sci. **269**(1498), 1383–1388 (2002)
4. De Masi, G., Prasetyo, J., Tuci, E., Ferrante, E.: Zealots attack and the revenge of the commons: quality vs quantity in the best-of-n. In: Proceedings of the 12th International Conference on Swarm Intelligence (ANTS), pp. 256–268 (2020)
5. De Masi, G., Prasetyo, J., Zakir, R., Mankovskii, N., Ferrante, E., Tuci, E.: Robot swarm democracy: the importance of informed individuals against zealots. Swarm Intell. J. (2021, inpress)
6. Dorigo, M., Şahin, E.: Guest editorial. Special issue: Swarm Robotics. Auton. Rob. **17**(2–3), 111–113 (2004)
7. Dudek, G., Jenkin, M.: Computational Principles of Mobile Robotics. Cambridge University Press, Cambridge (2000)
8. Ebert, J., Gauci, M., Nagpal, R.: Multi-feature collective decision making in robot swarms. In: Proceedings of the 17th International Conference on Autonomous Agents and MultiAgent Systems (AAMAS), pp. 1711–1719 (2018)
9. Kato, S., Jones, M.: An extended family of circular distributions related to wrapped Cauchy distributions via Brownian motion. Bernoulli **19**(1), 154–171 (2013)
10. Ligot, A., Birattari, M.: Simulation-only experiments to mimic the effects of the reality gap in the automatic design of robot swarms. Swarm Intell. **14**(1), 1–24 (2020)

11. Mondada, F., et al.: The e-puck, a robot designed for education in engineering. In: Proceedings of the 9th International Conference on Autonomous Robot Systems and Competitions, vol. 1, pp. 59–65 (2009)
12. Morlino, G., Trianni, V., Tuci, E.: Evolution of collective perception in a group of autonomous robots. In: Madani, K., Dourado Correia, A., Rosa, A., Filipe, J. (eds.) Computational Intelligence. SCI, vol. 399, pp. 67–80. Springer, Heidelberg (2012). https://doi.org/10.1007/978-3-642-27534-0_5
13. Scheidler, A., Brutschy, A., Ferrante, E., Dorigo, M.: The k -unanimity rule for self-organized decision-making in swarms of robots. IEEE Trans. Cybern. **46**, 1175–1188 (2016)
14. Strobel, V., Ferrer, E., Dorigo, M.: Managing byzantine robots via blockchain technology in a swarm robotics collective decision making scenario. In: Proceedings of the 17th International Conference on Autonomous Agents and MultiAgent Systems (AAMAS), pp. 541–549. International Foundation for Autonomous Agents and Multiagent Systems (2018)
15. Trianni, V., Tuci, E., Ampatzis, C., Dorigo, M.: Evolutionary swarm robotics: a theoretical and methodological itinerary from individual neuro-controllers to collective behaviour. In: Vargas, P., Di Paolo, E., Harvey, I., Husbands, P. (eds.) The Horizons of Evolutionary Robotics, pp. 153–178. MIT Press, Cambridge (2014)
16. Valentini, G., Brambilla, D., Hamann, H., Dorigo, M.: Collective perception of environmental features in a robot swarm. In: Dorigo, M., et al. (eds.) ANTS 2016. LNCS, vol. 9882, pp. 65–76. Springer, Cham (2016). https://doi.org/10.1007/978-3-319-44427-7_6
17. Valentini, G., Ferrante, E., Dorigo, M.: The best-of-n problem in robot swarms: formalization, state of the art, and novel perspectives. Front. Rob. AI **4**, 9 (2017). https://doi.org/10.3389/frobt.2017.00009
18. Valentini, G., Hamann, H., Dorigo, M.: Self-organized collective decision making: the weighted voter model. In: Proceedings of the 2014 International Conference on Autonomous Agents and Multi-Agent Systems (AAMAS), pp. 45–52. International Foundation for Autonomous Agents and Multiagent Systems (2014)
19. Valentini, G., Hamann, H., Dorigo, M.: Efficient decision-making in a self-organizing robot swarm: on the speed versus accuracy trade-off. In: Proceedings of the 2015 International Conference on Autonomous Agents and Multiagent Systems (AAMAS), pp. 1305–1314. International Foundation for Autonomous Agents and Multiagent Systems (2015)

A Novel Online Adaptation Mechanism in Artificial Systems Provides Phenotypic Plasticity

Michele Braccini[1,2]($\boxtimes$) (iD), Andrea Roli[1,2] (iD), and Stuart Kauffman[3] (iD)

[1] Department of Computer Science and Engineering (DISI),
Alma Mater Studiorum Università di Bologna, Cesena, Italy
m.braccini@unibo.it
[2] European Centre for Living Technology, Venice, Italy
[3] Institute for Systems Biology, Seattle, USA

Abstract. The ability to respond to environmental stimuli with appropriate actions is a property shared by all living organisms, and it is also sought in the design of robotic systems. Phenotypic plasticity provides a way for achieving this property as it characterises those organisms that, from one genotype, can express different phenotypes in response to different environments, without involving genetic modifications. In this work, we study phenotypic plasticity in robots that are equipped with online sensor adaptation. We show that Boolean network controlled robots can attain navigation with collision avoidance by adapting the coupling between proximity sensors and their controlling network without changing its structure. In other terms, these robots, while being characterised by one genotype (i.e. the network) can express a phenotype among many that are suited for the specific environment. We also show that the dynamical regime that makes it possible to attain the best overall performance is the critical one, bringing further evidence to the hypothesis that natural and artificial systems capable of optimally balancing robustness and adaptivity are critical.

1 Introduction

The Modern Synthesis fostered the idea that differential survival and reproduction success of biological organisms are responsibility of the genotype. Later, Mayr's and Waddington's works [31,45], paved the way for the formulation of a theoretical framework according' to which the phenotype—rather than the genotype—together with the environment and the development process play a primary role in the origin of novelty from an evolutionary perspective.

It is the dynamic of the complex interaction network among genes, and between genes and the environment, which drives the stable expression patterns and so ultimately affects the phenotype determination. Therefore, the ensemble of dynamics than can be generated by the genes composing the organism's

J. J. Schneider et al. (Eds.): WIVACE 2021, CCIS 1722, pp. 121–132, 2022.
https://doi.org/10.1007/978-3-031-23929-8_12

genetic code represents a source of diversification that can explain the birth of new phenotypes and, consequently, their affirmation on the evolutionary scale. In biology, the capacity of a genotype to produce different phenotypes depending on the environment in which it is located is defined as *phenotypic plasticity* [20,25]. Noteworthy is the hypothesis that phenotypic plasticity could allow, during the development, the crossing of valleys in the fitness landscape, which is not possible with evolution driven by mutations in the evolutionary scale as phenotypes in the valleys are less fit [35]. Therefore, even if mutations (random or not) contribute to the creation of diversification by modifying gene regulatory networks, they are not a necessary condition for phenotypic plasticity and assume a role of supporting actors. They are, however, implicated in the *genetic accommodation* process [3,36,46], or when a reorganisation of the genotype allows individuals of subsequent generations to achieve the same phenotype at a lower cost [5], in terms of time, resources, etc.

From a cybernetic standpoint [47], the adaptation process underpinned by phenotypic plasticity can be interpreted as the creation of an internal model [14] of the external world. This process makes it possible for an organism to compress the wealth of information coming from the external world into an internal representation that values only the information relevant for the organism's life: ranging from its survival, e.g., homeostasis by maintaining its *essential variables* within physiological ranges [1], to its function accomplishment. It is therefore of great importance to understand the fundamental elements of phenotypic plasticity and devise models that can be profitably used in the design of artificial systems, such as robots.

The aim of our work is twofold: for the cybernetic side, to propose a new mechanism for the on-line design of artificial agents that reproduces and, at the same time, exploits the property of phenotypic plasticity;[1] for biology, to promote the analysis of the implications of the results achieved in artificial contexts to improve biological models.

2   Creation of Novelty in Artificial Systems

Artificial devices have already proven to be able to give rise to the emergence of diversity and the creation of novelty. Pask conducted a remarkable experiment involving truly evolvable hardware [34]. The assembly developed its own sensors from scratch and therefore its own *relevance criteria* [13] from the outside world.

Inspired by Pask's works, Cariani proposes a classification of the kinds of adaptive behaviours attainable by physical devices [12]. According to it, only devices capable of constructing new hardware, and in particular sensors and effectors, can manipulate the kind of information needed to perform a given task and so create new semantic states [19]. As Cariani points out, this is analogous to the biological evolution of organs.

Although more abstract than Pask's work, attempts to evolve sensors into robotic agents are present in the literature [7,17,27,30]; however, they are

---

[1] Not necessarily mimicking development.

offline approaches. Few works apply some elements of evolution or adaptation in robotics in an online setting [10,11]. Anyway, none of them may be assimilated to a phenotypic plasticity mechanism.

3   The Proposed Mechanism

We believe that the relevance of phenotypic plasticity in the design of artificial agents has so far been underestimated; only recently its prospective role and the properties that promotes in the robotic field are emerging [15,22,23,32]. Taking inspiration from biology, phenotypic plasticity mechanisms applied on artificial entities can be seen as an *online adaptation process*.

We propose an *online* adaptive mechanism capable of giving rise to the observable phenotypic plasticity property typical of biological organisms without requiring mutations. The mechanism can be described in its more abstract and generic form by the flowchart schema in Fig. 1.

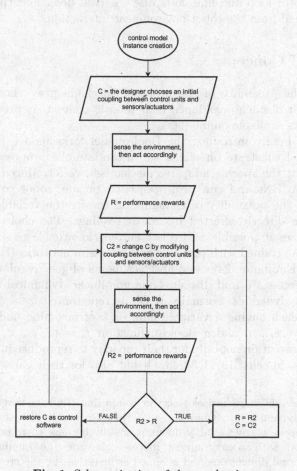

Fig. 1. Schematisation of the mechanism.

To be implemented, the mechanism requires a control model governing the artificial agent's behaviour that takes into account the possibility of modifying the coupling between the control units of the controller and the sensors or actuators, or both.

This adaptive mechanism is a case of (artificial) phenotypic plasticity because the adaptation:

- involves only the way external information is filtered by the artificial agents or how the internal information is exploited to produce coherent actions with the environment by acting on the internal sensing and actuation modules;
- takes place during the "life" of the individual and it is based on a feedback that rewards specific behaviours without changing the actual "genetic" structure.

Fundamentally, our mechanism mimics a kind of development of sensors/actuators tailored to the specific environment: the mapping between sensors readings and robot control program is the result of the embodied adaptation of the sensory-motor loop and manifests one observed behaviour (particular phenotype), emerged from the robot-environment interaction.[2]

4   Proof of Concept

To illustrate the feasibility of the proposed mechanism we chose the task of navigation with obstacle avoidance in a robotic context as proof of concept; code and results available online [9].

Firstly, we briefly introduce the Boolean networks model, since it represents the actual substrate on which we instantiate the proposed mechanism. We remark that the specific adaptive mechanism we introduced in not necessarily bound to BNs and can be instantiated on any robot control software accounting for the possibility of changing on the fly the variables of the controller that are directly affected by sensor readings. The choice of BN-based controllers makes it possible to exploit current knowledge on BNs and compare the results attained with previous ones. Boolean networks (BNs) have been introduced by Kauffman [24] as an abstract model of gene regulatory networks. They are discrete-state and discrete-time non-linear dynamical systems capable of complex dynamics. Formally, a BN is represented by a directed graph with n nodes each having a variable number k of incoming nodes, a Boolean variable x_i, $i = 1, ..., n$ and a Boolean function $f_i = (x_{i_1}, ..., x_{i_k})$. They have received much attention not only for their capacity to reproduce important properties of biological cells [16,21,38,43,44] but also for their capacity to express

---

[2] A similar principle characterises reservoir computing (RC), in that it also exploits the inherent dynamics of a dynamic system. However, besides this analogy the mechanism we propose is considerably different, mainly because there is no training set but interactions with an environment; moreover, it varies the couplings in both spatial and temporal dimensions and is aimed at investigating the creation of novelty and emergence of sensors.

rich dynamics combined with their relatively compact description, characteristics that make them appealing also for artificial applications. For this reason, the so-called Boolean network robotics takes advantage of BNs by employing them as control software in robots. Some examples of the remarkable results that could be obtained through their employment are reported in the following works [39–41].

The actual instantiation we propose—which is grounded on BN-robotics—consists of using a Boolean network as a robot program. Its dynamics, in a way similar to that performed by gene regulatory networks in biological cells [8], determines the behaviour of the robot and ultimately its final phenotype. In this context, with the word *phenotype* we identify the overall observable behaviour achieved by the artificial agent.

As illustrated in [42] the first step to take when designing a robot control software based on Boolean networks is to choose the coupling between nodes and robot's actuators and sensors: although it can be subject to (offline) optimisation driven by a fitness function, this mapping is usually chosen at design-time and stay the same during all the design.

Instead, in our approach the BN chosen as control software for the robotic agent is generated once and remains unchanged during all the robot life.[3] What distinguishes our approach from past ones is the fact that what changes is the coupling between the BN nodes and the sensors of the agents.[4] The coupling changes are not externally imposed by the designer: the robot determines which couplings are more suitable to it by continually interacting with the environment in which it is located.

As mentioned before, the task chosen for our proof of concept is that of navigation with obstacle avoidance. The robot, equipped with proximity sensors, must, therefore, try to move in its environment, represented by an arena and at the same time avoid colliding with the obstacles present in it. This problem can be formally considered as a dynamic classification problem. Indeed, the artificial agent is faced with a problem of classification of time series which are not independent of agent's behaviour since they are conditioned by the dynamics generated by its *sensorimotor loop* [28].

As a learning feedback, we provide a merit figure for assessing the degree of adaptation attained by the robot. This objective function will be discussed in Sect. 4.1.

Albeit in a more abstract form, this mechanism takes inspiration from Pask's evolvable and self-organising device. Here, the space of possibilities among which the robot can choose it's not open-ended, like the one used in Pask's experiment, but it is limited by the possible set of dynamics of the Boolean network, the number of sensors and the coupling combinations between the two. Simultaneously, it can be considered an artificial counterpart of the adaptive behaviour without

---

[3] In the present discussion, we will refer only to robots with fixed morphology, although this mechanism finds natural application in self-assembling robots.

[4] We remark that mechanism is abstract enough to be able to contemplate the variations of both sensors and actuators.

Fig. 2. The arena used in the experiments.

mutations present in the development phases of biological organisms. Indeed, the robot exploits the feedbacks it receives from the environment and consequently tries to improve the way it uses the dynamics provided by its own *raw genetic material*, i.e. the BN. In doing so, it does not modify BN functions nor topology but it uses the intrinsic BN's information processing capabilities.

In addition to testing the proposed mechanism we want to start investigating which general principles govern the best performing robotic agents, and therefore they promote and at the same time take advantage of the phenotypic plasticity characteristic of our adaptive mechanism. Fortunately, the literature relating to Boolean networks provides us with a wide range of both theoretical and experimental results to start from. A natural starting point for an analysis of differential performances obtained through the use of Boolean network models is that concerning the dynamical regimes in which they operate. So, in the next sections we investigate what Boolean network dynamical regimes—ordered, critical or chaotic—provides an advantage for robots equipped with the adaptive mechanism we have just introduced.

4.1 Experimental Setting

In our experiments we used a *foot-bot* robot model equipped with 24 proximity sensors (evenly placed along its main circumference) and controlled by two motorised wheels. The robot moves inside a squared arena, delimited by walls, with a central box (see Fig. 2). The goal we want the robot to achieve is to move as fast as possible around the central box without colliding against walls and the box itself. The robot is controlled by a BN. The coupling between the BN and the robot is as follows: two nodes are randomly chosen and their value is taken to control the two motors, which can be either ON (node with value 1) or OFF (node with value 0) and control the wheels at constant speed. The sensor readings return a value in $[0, 1]$ and so are binarised through a step function with threshold θ:[5] if the sensor value is greater than θ, then the BN node is set to 1, otherwise it is set to 0. The 24 sensors are randomly associated to 24 randomly chosen nodes in the network, excluding the output ones. At each network update, the binarised values from the sensors are overridden to the current values of the corresponding nodes, so as to provide an external signal to the BN.

The adaptive mechanism consists in randomly rewiring q connections between sensors and BN nodes (excluding output nodes, of course). The actual value of q is randomly chosen at each iteration in $\{1, 2, \ldots, 6\}$. The robot is then run for $T = 1200$ steps (corresponding to 120 seconds of real time); if the current binding enables the robot to perform better, or anyway the performance is not worse,[6] then it is kept, otherwise it is rejected and the previous one is taken as the basis for a new perturbation. Observe that this choice is anyway influenced by the experience of the robot accumulated in the last trial period and might also depend on sensor and actuator noise. We remark that the binding between proximity sensors and BN "input" nodes is the only change made to the network: in this way we address the question as to what extent a random BN can indeed provide a sufficient bouquet of behaviours to enable a robot to adapt to a given (minimally cognitive) task.

BNs are generated with n nodes, $k = 3$ inputs per node and random Boolean functions defined by means of the bias b, i.e. b is the probability of assigning a 1 a truth table entry. In the experiments we tested $n \in \{100, 1000\}$ and $b \in \{0.1, 0.21, 0.5, 0.79, 0.9\}$. According to [29], random BNs with $k = 3$ generated with bias equal to 0.1 or 0.9 are likely to be ordered, with bias equal to 0.5 are likely to be chaotic and bias equal to 0.21 and 0.79 characterises criticality.[7] Only the BN nodes controlling the wheels have function randomly chosen always with bias 0.5; this is to avoid naively conditioning the behaviour of the robot, which would tend to be always moving (resp. resting) for high biases (resp. low biases). This choice has anyway a negligible contribution to the overall dynamical regime of the network.

The performance is evaluated by an objective function that is accumulated along the robot execution steps and then normalised. The function is defined as follows:

$$F = (1 - p_{max}) \left(1 - |v_l - v_r|\right) \frac{(v_l + v_r)}{2}$$

where p_{max} is the maximal value returned among the proximity sensors, and v_l and v_r are the binarised values used to control the left and right motor, respectively. The intuition of the function is to favour fast and as much straight as possible trajectories far from the obstacles [33]. Experiments are run in simulations with ARGoS [37].

---

[5] In our experiments we set $\theta = 0.1$.

[6] Accepting changes that do not deteriorate the performance is helpful because it enables a wider exploration of the space of possible configurations.

[7] Along the critical line, k and b are linked by this relation: $k = \dfrac{1}{2b(1 - b)}$ [4].

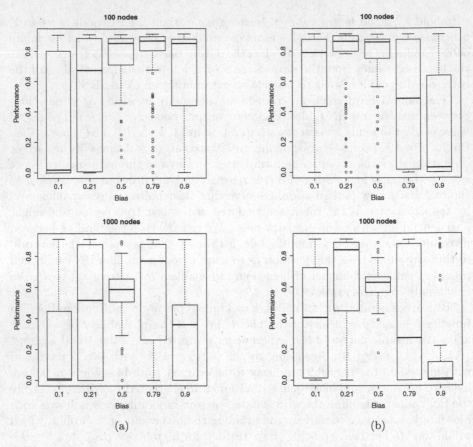

Fig. 3. Boxplots summarising the performance as a function of BN bias for BNs with $n = 100$ and $n = 1000$. On the left side (a) the results concerning the "normal" encoding are reported, while on the right side (b) the results for robots controlled by BNs with a "dual" encoding (i.e., 0 denotes that an obstacle is detected) are outlined.

4.2   Results

We run 1000 random replicas for each configuration of BN parameters and collected statistics on the best performance attained after a trial of 1.44×10^4 seconds (corresponding to 1.44×10^5 steps in total). In order to avoid variance due to the initialisation of the network, all nodes are initially set to 0. Since the evaluation function can not be maximal across the whole run, as the robot must anyway turn to remain in the arena and avoid the obstacles, values of F greater than 0.7 correspond to a good performance. As we can observe in Fig. 3a, despite the simple adaptation mechanism, a large fraction of BN attains a good performance. Notably, critical networks attains the best performance—this result is striking for large BNs ($n = 1000$). In particular, for $n = 100$ the results achieved with $b = 0.21$ are significantly better (Wilcoxon test, with $\alpha = 0.05$) than all the other cases, with the exception of $b = 0.5$ for which we could not reject the null hypothesis. As for $n = 1000$ the case with $b = 0.21$ is significantly better than all the other ones. We observe, however, that just one of the two bias values corresponding to the critical regime corresponds to a good performance. The reason is that in our experiment the symmetry between 0 and 1 is broken, because a 1 means that an obstacle is detected. To test the robot in the dual condition, we ran the same experiments with a negative convention on the values (if the obstacle is near, then the node is set to 0; similarly, the wheels are activated if the corresponding output node state is 0). As expected, results (see Fig. 3b) are perfectly specular to the previous ones (and the same results of the statistical test hold).

The picture emerging from our experiments is neat: one bias value characterises the best overall performance and this value is one of the two along the critical line. The reason of the asymmetry between the two critical bias values has to be ascribed to the symmetry breaking introduced by the binarisation of the sensor values. Anyway, the remarkable observation is that random BNs generated with a bias corresponding to critical regime adapt better than the other kinds of BNs. Since the adaptive mechanism only acts on the mapping between sensors and input nodes, the dynamical regime of the BNs is preserved; therefore, we have a further evidence that critical BNs achieve the best performance in discriminating the external signals.

5   Conclusion

In this work we have introduced a novel mechanism for the on-line design of artificial agents inspired by the biological property of phenotypic plasticity. Besides proving its effectiveness in a simple but relevant task for the robotic literature, i.e. navigation with obstacle avoidance, we also observed that robots controlled by *critical* BNs achieve the highest level of performance.

The online sensor adaptation mechanism introduced in this work consists in varying the coupling between robot sensors and BN nodes. Other possible adaptive mechanisms are subject of ongoing work and preliminary results, involving also more complex scenarios, e.g. swarm of robots engaged in a foraging task, suggest that sensor and actuator adaptation mechanisms are way better than those concerning structural modifications of the BN, and that critical BNs again attain superior performance. To reveal all the potential of this adaptive mechanism it will be necessary to test it in *i)* complex adaptive scenarios that assume the capacity of switch between different tasks in dynamic and changing environments; *ii)* learning procedures involving morphological and environmental scaffolding [6]. Indeed, we believe that phenotypic plasticity can be advantageous and apparent when changes in the morphology are involved too. In addition, we plan to investigate the relation between criticality of controlling BNs, their performance and the maximisation of some information theory measures, such as predictive information [2], integrated information [18] and transfer entropy [26].

Besides providing evidence to the *criticality hypothesis*, the results we have presented make it possible to speculate further: criticality may be a property that enables phenotypic plasticity—at least as long as sensory adaptation is concerned. We believe that this outcome provides a motivation for deeper investigations, which may be primarily conducted in simulation or anyway with artificial systems. Nevertheless, we also envisage the possibility of devising wet experiments, in which the dynamical regime of an organism is externally controlled and its ability to exhibit phenotypic plasticity can be estimated.

In addition, a mechanism like the one we have introduced may be an effective tool for tuning artificial systems to the specific environment in which they have to operate.

References

1. Ashby, W.: Design for a Brain. Chapman & Hall, Boca Raton (1952)
2. Ay, N., Bertschinger, N., Der, R., Güttler, F., Olbrich, E.: Predictive information and explorative behavior of autonomous robots. Eur. Phys. J. B - Condensed Matter Compl. Syst. **63**(3), 329–339 (2008)
3. Baldwin, J.: A new factor in evolution. Am. Natural. **30**(355), 536–553 (1896)
4. Bastolla, U., Parisi, G.: A numerical study of the critical line of Kauffman networks. J. Theor. Biol. **187**(1), 117–133 (1997)
5. Bateson, P., Gluckman, P.: Plasticity, Robustness, Development and Evolution. Cambridge University Press, Cambridge (2011)
6. Bongard, J.C.: Morphological and environmental scaffolding synergize when evolving robot controllers: artificial life/robotics/evolvable hardware. In: Proceedings of the 13th Annual Conference on Genetic and Evolutionary Computation. GECCO 2011, pp. 179–186. Association for Computing Machinery, New York (2011)
7. Bongard, J.C., Pfeifer, R.: Evolving complete agents using artificial ontogeny. In: Hara, F., Pfeifer, R. (eds.) Morpho-functional Machines: The New Species, pp. 237–258. Springer, Tokyo (2003). https://doi.org/10.1007/978-4-431-67869-4_12
8. Braccini, M.: Applications of biological cell models in robotics. arXiv preprint arXiv:1712.02303 (2017)
9. Braccini, M., Roli, A., Kauffman, S.A.: Release of code and results of the paper "a novel online adaptation mechanism in artificial systems provides phenotypic plasticity" (v1.0.0). Zenodo (2021). https://doi.org/10.5281/zenodo.4811251
10. Brawer, J., Hill, A., Livingston, K., Aaron, E., Bongard, J., Long, J.: epigenetic operators and the evolution of physically embodied robots. Front. Rob. AI **4**, 1 (2017)
11. Bredeche, N., Haasdijk, E., Prieto, A.: Embodied evolution in collective robotics: a review. Front. Rob. AI **5**, 12 (2018)
12. Cariani, P.: Some epistemological implications of devices which construct their own sensors and effectors. In: Toward a Practice of Autonomous Systems: Proceedings of the First European Conference on Artificial Life, pp. 484–493. MIT Press (1992)
13. Cariani, P.: To evolve an ear. Epistemological implications of Gordon Pask's electrochemical devices. Syst. Res. Behav. Sci. **10**(3), 19–33 (1993)
14. Conant, R.C., Ashby, W.R.: Every good regulator of a system must be a model of that system. Int. J. Syst. Sci. **1**(2), 89–97 (1970)
15. Damper, R.I., Elliott, T., Shadbolt, N.R.: Developmental robotics: manifesto and application. Philos. Trans. Roy. Soc. Lond. Ser. A: Math. Phys. Eng. Sci. **361**(1811), 2187–2206 (2003)

16. Daniels, B.C., et al.: Criticality distinguishes the ensemble of biological regulatory networks. Phys. Rev. Lett. **121**, 138102 (2018)
17. Dellaert, F., Beer, R.: A developmental model for the evolution of complete autonomous agents. In: Proceedings of SAB 1996, January 1996
18. Edlund, J., Chaumont, N., Hintze, A., Koch, C., Tononi, G., Adami, C.: Integrated information increases with fitness in the evolution of animates. PLoS Comput. Biol. **7**(10), e1002236 (2011)
19. Emmeche, C., Kull, K.: Towards a Semiotic Biology: Life is the Action of Signs. World Scientific, Singapore (2011)
20. Fusco, G., Minelli, A.: Phenotypic plasticity in development and evolution: facts and concepts. Philos. Trans. Roy. Soc. B: Biol. Sci. **365**(1540), 547–556 (2010)
21. Helikar, T., et al.: The cell collective: toward an open and collaborative approach to systems biology. BMC Syst. Biol. **6**(1), 96 (2012)
22. Hunt, E.R.: Phenotypic plasticity as a framework of bio-inspiration for minimal field swarm robotics. Front. Rob. AI **7** (2020)
23. Jin, Y., Meng, Y.: Morphogenetic robotics: an emerging new field in developmental robotics. IEEE Trans. Syst. Man Cybern. Part C: Appl. Rev. **41**(2), 145–160 (2011)
24. Kauffman, S.: Metabolic stability and epigenesis in randomly constructed genetic nets. J. Theor. Biol. **22**(3), 437–467 (1969)
25. Kelly, S., Panhuis, T., Stoehr, A.: Phenotypic plasticity: molecular mechanisms and adaptive significance. Compr. Physiol. **2**(2), 1417–1439 (2011)
26. Lizier, J., Prokopenko, M., Zomaya, A.: The information dynamics of phase transitions in random Boolean networks. In: ALIFE, pp. 374–381 (2008)
27. Lund, H., Hallam, J., Lee, W.-P.: Evolving robot morphology. In: Proceedings of 1997 IEEE International Conference on Evolutionary Computation (ICEC 1997), pp. 197–202, April 1997
28. Lungarella, M., Pfeifer, R.: Robots as cognitive tools: information theoretic analysis of sensory-motor data. In: EDEC 2001-Symposium at the International Conference on Cognitive Science (2001)
29. Luque, B., Solé, R.: Phase transitions in random networks: simple analytic determination of critical points. Phys. Rev. E **55**(1) (1997)
30. Mark, A., Mark, R., Polani, D., Uthmann, T.: A framework for sensor evolution in a population of Braitenberg vehicle-like agents. In: Adami, C., Belew, R.K., Kitano, H., Taylor, C.E. (eds.) Sixth International Conference on Artificial Life (ALIFE VI), pp. 428–432. MIT Press (1998)
31. Mayr, E.: The objects of selection. Proc. Natl. Acad. Sci. **94**(6), 2091–2094 (1997)
32. Miras, K., Ferrante, E., Eiben, A.E.: Environmental regulation using plasticoding for the evolution of robots. Front. Rob. AI **7**, 107 (2020)
33. Nolfi, S., Floreano, D.: Evolutionary Robotics. The MIT Press, Cambridge (2000)
34. Pask, G., et al.: Physical analogues to the growth of a concept. In: Mechanization of Thought Processes, Symposium, vol. 10, pp. 765–794 (1958)
35. Pfennig, D.W., Wund, M.A., Snell-Rood, E.C., Cruickshank, T., Schlichting, C.D., Moczek, A.P.: Phenotypic plasticity's impacts on diversification and speciation. Trends Ecol. Evol. **25**(8), 459–467 (2010)
36. Pigliucci, M., Murren, C., Schlichting, C.: Phenotypic plasticity and evolution by genetic assimilation. J. Exp. Biol. **209**(12), 2362–2367 (2006)
37. Pinciroli, C., et al.: ARGoS: a modular, multi-engine simulator for heterogeneous swarm robotics. Swarm Intell. **6**(4), 271–295 (2012)
38. Roli, A., Villani, M., Filisetti, A., Serra, R.: Dynamical criticality: overview and open questions. J. Syst. Sci. Complex. **31**(3), 647–663 (2018)

39. Roli, A., Villani, M., Serra, R., Benedettini, S., Pinciroli, C., Birattari, M.: Dynamical properties of artificially evolved Boolean network robots. In: Gavanelli, M., Lamma, E., Riguzzi, F. (eds.) AI*IA 2015. LNCS (LNAI), vol. 9336, pp. 45–57. Springer, Cham (2015). https://doi.org/10.1007/978-3-319-24309-2_4

40. Roli, A., Benedettini, S., Birattari, M., Pinciroli, C., Serra, R., Villani, M.: A preliminary study on BN-robots' dynamics. In: Proceedings of the Italian Workshop on Artificial Life and Evolutionary Computation (WIVACE 2012), pp. 1–4 (2012)

41. Roli, A., Braccini, M.: Attractor landscape: a bridge between robotics and synthetic biology. Complex Syst. **27**, 229–248 (2018)

42. Roli, A., Manfroni, M., Pinciroli, C., Birattari, M.: On the design of Boolean network robots. In: Di Chio, C., et al. (eds.) EvoApplications 2011. LNCS, vol. 6624, pp. 43–52. Springer, Heidelberg (2011). https://doi.org/10.1007/978-3-642-20525-5_5

43. Serra, R., Villani, M., Graudenzi, A., Kauffman, S.A.: On the distribution of small avalanches in random Boolean networks. In: Ruusovori, P., et al. (eds.) Proceedings of the 4th TICSP Workshop on Computational Systems Biology, pp. 93–96. Juvenes print, Tampere (2006)

44. Shmulevich, I., Kauffman, S., Aldana, M.: Eukaryotic cells are dynamically ordered or critical but not chaotic. Proc. Natl. Acad. Sci. **102**(38), 13439–13444 (2005)

45. Waddington, C.H.: The Epigenotype. Endeavour (1942)

46. West-Eberhard, M.J.: Developmental plasticity and the origin of species differences. Proc. Natl. Acad. Sci. **102**(suppl 1), 6543–6549 (2005)

47. Wiener, N.: Cybernetics: Or Control and Communication in the Animal and the Machine. MIT Press, Cambridge (1948)

Computer Vision and Computational Creativity

Exploration of Genetic Algorithms and CNN for Melanoma Classification

Luigi Di Biasi[1] , Alessia Auriemma Citarella[1]([✉]) , Fabiola De Marco[1] ,
Michele Risi[1] , Genoveffa Tortora[1] , and Stefano Piotto[2]

[1] Department of Computer Science, University of Salerno,
Via Giovanni Paolo II, 132, 84084 Fisciano, Italy
aauriemmacitarella@unisa.it
[2] Department of Pharmacy, University of Salerno,
Via Giovanni Paolo II, 132, 84084 Fisciano, Italy

Abstract. This contribution reports the first results we achieved by
combining the main capabilities of Genetic Algorithms (GA) with Convolutional Neural Network (CNN) to address the melanoma detection
problem (GACNN). The presented results are related to the melanoma
classification problem we chose due to the large proposal available in literature usable for performance comparison. We used a clinical dataset (i.e.,
MED-NODE) as the data source. We compared performance obtained
by GACNN with the AlexNet both with and without the Otsu segmentation. In addition, we considered the *accuracy* parameter as the scoring
function for the GA evolution process. The preliminary results suggest
the proposed approach could improve melanoma classification by allowing network design to evolve without user interaction. Furthermore, the
suggested approach can be extended to additional melanoma datasets
(e.g., clinical, dermoscopic, or histological) in the future with other innovative evolutionary optimization algorithms.

Keywords: Melanoma detection · Genetic algorithms · Neural
networks

1 Introduction

Skin cancer is one of the most dangerous and deadly cancers. Unfortunately, the
incidence of skin cancer has been rising in recent years and, for some subtypes,
the biggest problem is a lack of early detection, a limiting issue for first-line
therapy in cases of this malignant pathology [5]. In the latest years, new competitions, such as ISIC[1], and new melanoma detection tools, implemented as
a statistical tool, machine/deep learning software or expert system, and techniques were proposed. These tools can work on multiple data-source: clinical,
dermoscopic, or histological images can be used. Other tools might use analytic
data extracted from the Electronic Health Record or anamnesis of a patient. For

---

[1] https://challenge.isic-archive.com/.

J. J. Schneider et al. (Eds.): WIVACE 2021, CCIS 1722, pp. 135–138, 2022.
https://doi.org/10.1007/978-3-031-23929-8_13

image analysis, CNN or similar are often used, but the research is not limited to a single Neural Network (NN) architecture. In this work, we want to present the preliminary results we obtained by merging GAs core functions (selection, mutation, and cross-over) with what we are defining as an Evolutionary-based CNN design approach. Our working hypothesis states that an NN population that uses the GA approach can converge to an acceptable solution (high accuracy in prediction and confusion matrices), limiting the NN growth of layers. In particular, we do not aim to use GA to improve the determination of hyperparameters on a defined (and static) NN. Instead, we aim to design a self-assembly NN population driven by its performance regarding a well-defined problem.

2   Related Works

Genetic Algorithms are inspired by the theory of biological evolution and are used for many process optimizations. The algorithm starts with a random population of network designs and then iterates through three stages: selection, crossover, and mutation [3]. Although it is possible to retrieve attempts to combine neural networks and evolutionary algorithms still from 1990 [7], the computation efforts needed to merge these techniques became more tractable only in the latest years thanks to the major cloud provider that permits to loan high computational architecture without the needed to build it from scratch. Many research in various fields have used a combination of neural networks and genetic algorithms to solve optimization and classification problems over the last two decades, ranging from river water quality prediction [1] to the most recent tuning of numerous hyper-parameters simultaneously [4].

3   Materials and Methods

In our experimentation environment (`Matlab 2021`), we defined our working objects following the GA terminology. Where used, the notation $F(t)$ indicates an object's composition at the time t. In particular, we define an entity E_i as a vector $E_i = \{F_1, \ldots, F_m\}$ of m features. We called each feature F_j of a generic entity E_i a gene of E_i. The entire set of genes is called the Genome of E_i. In our simulation, each gene can represent a Matlab CNN core object (network layer) or a pre-processing routine, in particular Otsu segmentation [6]. Each feature F_j can be expressed or not by E_i. This means that a new entity E_k could inherit a gene F_e from E_i starting to express it. Then, in our simulation, we have a silent and expressed genes. The set $P(t) = \{E_1, \ldots, E_n\}$ is called Population at time t. The population size $n(t)$ might vary related to t. We defined the following constraints: the first gene of each entity must be an image input (II) or one of the pre-processing routines we defined before; if the gene q is a pre-processing routine, then the gene $g + 1$ must be a II layer or another pre-processing layer; the latest gene of an entity must be a classification layer.

In our experimentation, the population is the set of all live entities. For the experimentation, we limited the gene type that an entity can use to: *Convolution,*

ReLu, Cross Channel Normalization, Max Pooling Grouped Convolution, Fully Connected Layer, Dropout and *Softmax*. In each evolution step, all the entities with expressed genes not compatible with the environment die immediately. Then, if an entity exposes a genes pipeline that is not allowed by Matlab *training* function, it dies immediately. For example, if the first gene of the entity E_i is a II layer with input dimension $D = Width \times Height \times Depth$ and the second gene is a C layer, it must use the same D input size. Otherwise, the training function will crash, and we considered the expressed gene is not compatible with the simulated environment (melanoma classification).

Each compatible entity was trained using the following configuration: 'sgdm', 'MaxEpochs', 16, 'MiniBatchSize', 12, 'Shuffle', 'every-epoch', 'InitialLe-arnRate', 0.0001, 'ExecutionEnvironment', 'auto'.

We used the maximization of the global population accuracy as the function to drive the population evolution. For each evolutionary step, we computed the maximum accuracy from each survived entity. Thus, for each generation, all the entity that exposes an accuracy at time t equals or better than the previous generation $t - 1$ maximum accuracy will survive. Also, a random 10% of entities were chosen at each step to survive independently regarding the accuracy at time t. If no accuracy improvement was detected for ten consecutive evolution steps, the GA was stopped. Due to the physical limits of our cloud platform, we were forced to limit the possible crossover and mutation. As a result, only ten mutations and 100 crossovers were allowed for each survived entity. We tried to mitigate these constraints with an initial randomized population of 10K entities. We forced evolution to stop after 100 iterations. The dataset used is MED-NODE, containing a total of 170 clinical images, specifically 70 images of melanoma and 100 of benign nevi [2].

4   Results and Discussion

We executed the AlexNet network with and without Otsu segmentation applied to MED-NODE, repeating the training step 100 times. Then, we executed the GACNN allowing the system to evolve for 100 iterations. As a reference parameter, we used *Accuracy* (ACC). For the standard AlexNet execution, we computed average ACC (*mean ACC*), maximum ACC (*max ACC*), minimum ACC (*min ACC*) and Standard Deviation (*SD*), as reported in Table 1. The best AlexNet performance was 0.97%, while the mean ACC was 0.81%. For GACCN, we evaluated the ACC trend over the evolution steps as reported in Fig. 1. The max ACC reached by GACNN is 0.97 before reaching the 100th iteration.

Table 1. AlexNet Performances on the MED-NODE dataset.

MED-NODE					
Net	Segmentation	min ACC	max ACC	mean ACC	SD
AlexNet	–	0.68	0.97	0.81	0.06
	Otsu	0.50	0.91	0.72	0.07
GACNN	–	0.68	0.97	–	–

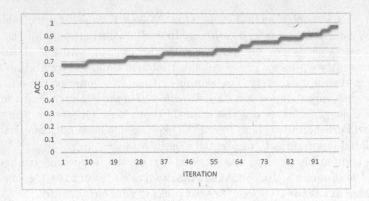

Fig. 1. GACNN performance over 100 iterations.

5   Conclusion

The preliminary results suggest that allowing GA to define NN structure design might permit performance comparable to the classic NN training methods. In particular, GACNN outperforms AlexNet mean ACC over 100 executions. Figure 1 also depicts a plateau set that shows a reduction trend every nine iterations in mean. These results might suggest the convergence of the population to the optimal solution. In contrast, we observed a high death ratio (up to 95% for each evolution step). This fact might suggest a need for a better definition of initial population or recombination steps. A more deeper study is needed to analyze population evolution and behaviors, in particular the death ratio.

References

1. Ding, Y., Cai, Y., Sun, P., Chen, B.: The use of combined neural networks and genetic algorithms for prediction of river water quality. J. Appl. Res. Technol. **12**(3), 493–499 (2014)
2. Giotis, I., Molders, N., Land, S., Biehl, M., Jonkman, M.F., Petkov, N.: MED-NODE: a computer-assisted melanoma diagnosis system using non-dermoscopic images. Expert Syst. Appl. **42**(19), 6578–6585 (2015)
3. Kramer, O.: Genetic algorithms. In: Kramer, O. (ed.) Genetic Algorithm Essentials. SCI, vol. 679, pp. 11–19. Springer, Cham (2017). https://doi.org/10.1007/978-3-319-52156-5_2
4. Kumar, P., Batra, S., Raman, B.: Deep neural network hyper-parameter tuning through twofold genetic approach. Soft. Comput. **25**(13), 8747–8771 (2021). https://doi.org/10.1007/s00500-021-05770-w
5. Naik, P.P.: Cutaneous malignant melanoma: a review of early diagnosis and management. World J. Oncol. **12**(1), 7 (2021)
6. Qi, L.N., Zhang, B., Wang, Z.K.: Application of the Otsu method in image processing. Radio Eng. China **7**(009) (2006)
7. Whitley, D., Starkweather, T., Bogart, C.: Genetic algorithms and neural networks: optimizing connections and connectivity. Parallel Comput. **14**(3), 347–361 (1990)

Using Genetic Algorithms to Optimize a Deep Learning Based System for the Prediction of Cognitive Impairments

Nicole Dalia Cilia, Tiziana D'Alessandro, Claudio De Stefano, Francesco Fontanella[✉], and Alessandra Scotto di Freca

Department of Electrical and Information Engineering (DIEI), Universitá di Cassino e del Lazio Meridionale, Via G. Di Biasio, 43, 03043 Cassino, Frosinone, Italy
{nicoledalia.cilia,tiziana.dalessandro,destefano,
fontanella,a.scotto}@unicas.it

Abstract. Cognitive impairments are one of the first signs of neurodegenerative diseases, e.g. Alzheimer's and Parkinson's. For this reason, their early diagnosis is crucial to better manage the course of these diseases. Amongst the others, cognitive impairments also affect handwriting skills. In a previous research, we presented a study in which the handwriting movements of the participants were recorded while they were performing some elementary handwriting tasks. The acquired data were used to generate synthetic images in order to feed four different models of convolutional neural networks and collect the predictions and their confidence degrees. In this paper, we present a system that uses a Genetic Algorithm (GA) to improve the performance of the system previously presented. The genetic algorithm finds the subset of tasks that allows improving the prediction ability of the previously presented system. The experimental results confirmed the effectiveness of the proposed approach.

1 Introduction

Cognitive impairments (CI in the following) are defined as a condition in which a person experiences a slight, but noticeable, decline in mental abilities compared with others of the same age. In many cases CI evolve to Alzheimer's disease (AD). AD incidence increases strongly with age, and is expected that in the next decades it will dramatically increase. Unfortunately, there is no cure for AD, but its effects can only be slowed down with proper therapies. For this reason, an early diagnosis of CI is crucial. This creates a strong need for the improvement of the approaches currently being used for the diagnosis of these diseases.

Handwriting is one the skills compromised by CI [3, 19, 20, 25]. Indeed, it has been observed that the handwriting of cognitively impaired people is altered [21, 22]. In the last decade, many studies have investigated the use of machine

J. J. Schneider et al. (Eds.): WIVACE 2021, CCIS 1722, pp. 139–150, 2022.
https://doi.org/10.1007/978-3-031-23929-8_14

learning based techniques to analyze people's handwriting to detect those affected by CI. However, most of these involved a few dozens of people, thus limiting the effectiveness of the classification algorithms used. To try to overcome these problems, in [6] the authors proposed a protocol consisting of 25 handwriting tasks, with the aim of investigating how CI affect cognitive skills involved in the handwriting process. Currently, the protocol has been used to collect data from 181 participants, 90 cognitive impaired people (PT in the following) and a healthy control group made of 91 people (HC). This data contains both on-paper and in-air pen traits. The first represent the movements of the pen when its tip s touching the paper, whereas the second are recorded when the pen tip is in the air. This choice has been motivated by the fact that many studies have found that in-air traits analysis allows a better characterization of handwriting anomalies caused by CI [7, 8, 14, 16, 17]. The data collected was used to generate tiff images. The four channels of these images encoded the dynamic information extracted from the handwriting traits recorded, namely velocity, jerk, pressure, and acceleration. These images were used to train three different models of Convolutional Neural Networks (CNNs), on each of the 34 task of the protocol. Since we had 34 predictions for each participant (one for each task), the final prediction for each participant (Cognitively impaired or healthy) was achieved combining them using the Majority-vote as a combination rule.

In this paper, we present a GA-based system for the improvement of the prediction performance of the approach described above. In particular, the devised system selects the subset of tasks that allows improving the performance in assessing whether the involved subjects are affected by cognitive impairments or not. The final prediction of the cognitive state of a subject is made applying the majority vote rule to the responses from the selected tasks. The system was trained by using a dataset where each sample contains the responses provided by the CNNs for the total number of tasks. A part of this dataset was used as a training set, to implement the fitness function for evaluating the individuals to be evolved, whereas the remaining part was used as a test set to assess the performance of the system on unseen data.

We tested our approach taking into account three well-known and commonly used CNNs: VGG19, ResNet50 and InceptionV3. Moreover, in the first set of experiments, we analyzed the generalization ability of the system as well as its capability of reducing the number of tasks needed to correctly predict cognitive impairments. Then, by counting the occurrences of the tasks in the solutions found in the several runs performed, we tried to figure out the most relevant tasks.

The remainder of the paper is organized as follows: Sect. 2 shows the last researches related to this work. Section 3 describes the data collection and the protocol developed to collect handwriting data. Section 4 details the proposed system. Section 5 describes the experiments and presents the results achieved. Finally, Sect. 6 is devoted to some concluding remarks.

2   Related Work

Evolutionary algorithms have proved to be effective search tools in solving many real-world problems characterized by large and non-linear search spaces [5,10,12], and have been widely used also for health applications, especially GA and GP. For example, in [2], the authors proposed a constrained-syntax GP-based algorithm for discovering classification rules in medical data. The approach proposed was tested on several datasets and achieved better results than decision trees. In [4], the authors used GP to solve a problem involving the prediction of physio-chemical properties of proteins in tertiary structure. The authors proved that the proposed approach was more effective than Artificial Neural Networks and Support Vector Machines. More recently, the Cartesian GP approach has been used to automatically identify subjects affected by the Parkinson's disease through the analysis of their handwriting [23,24]. GP algorithms have been also used as a tool to support medical decisions for treating rare diseases [1].

Also GA has been widely used in medical applications, in most of the medical specialities, e.g., medical imaging, rehabilitation medicine, and health care management [15]. In [26] the authors used a GA with Multi-Objective fitness function to find the relevant volumes of the brain affected by AD, whereas in [18] the authors used a GA to search the optimal set of neuropsychological tests, to include in a system for the prediction of AD.

3   Data Collection

The 181 participants (90 cognitively impaired patients and 91 healthy people) were recruited with the support of the geriatric ward, Alzheimer unit, of the University hospital Federico II in Naples. We used clinical tests (such as PET, TAC and enzymatic analysis) and standard cognitive tests (such as MMSE) to define participants'recruiting criteria. In order to reduce the data bias as much as possible, demographic and educational characteristics of the control group were matched with those of the patient group.

We collected data by means of a graphic tablet, which allowed participants to write on standard A4 white sheets using a biro pen. Such a pen produced both the ink trace on the sheet and spatial coordinates (x,y) as well as the pressure of the participant handwriting movements. These coordinates were acquired at frequency of 200 Hz. The tablet also recorded the in-air movements (up to a maximum of three centimetres from the sheet). Note that this experimental setting allowed participants to keep their movements as natural as possible.

The aim of the protocol used to acquire participants handwriting data is to investigate whether there are specific features and/or tasks that allow us to better distinguish cognitively impaired people from the control group. The goal of these tasks is to test the patients' capability of repeating simple as well as complex graphic gestures, which have a semantic meaning, such as letters and words of different lengths and with different spatial organizations. The tasks

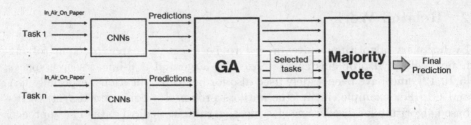

Fig. 1. The layout of the proposed system. Note that in our case $n = 34$.

are arranged in increasing order of difficulty, in terms of the cognitive functions required [6] and are grouped as follows:

- *Graphic* tasks: they test the patient's ability in: (i) writing elementary traits; (ii) joining some points; (iii) drawing figures (simple or complex and scaled in various dimensions).
- *Copy* and *Reverse Copy* tasks: they assess patient's capabilities in repeating complex graphic gestures, with semantic meaning (letters, words and numbers of different lengths and with different spatial organizations).
- *Memory* tasks: they test the variation of the graphic section, keeping in memory a word, a letter, a graphic gesture or a motor planning.
- *Dictation* tasks: they allow us to investigate how the writing in the task varies (with phrases or numbers) in which the use of the working memory is necessary.

The protocol consists of 25 tasks, but from these 9 more tasks are obtained by splitting the single words from the tasks in which was asked the subject to write a list of different words (task 14 and 18). Thus, the whole dataset is composed of 34 tasks.

4   The Proposed System

The following subsections detail the steps performed to implement the proposed system (see Fig. 1).

4.1   Image Generation

Starting from the raw coordinates acquired during the protocol execution, $\mathbf{x}$, $\mathbf{y}$, we have generated a dataset of images to be submitted to the CNN networks. The traits of the synthetic images are obtained by considering the points (x_i, y_i) as vertices of the polygonal that approximates the original curve, so every trait is delimited by the couple of points (x_i, y_i) and (x_{i+1}, y_{i+1}). As the tablet also store in air movements, the generated images contain not only the on-paper traits, but also those in air, for these reasons they are called In-Air_On-Paper images. We created multi-channels TIFF images, storing four representations

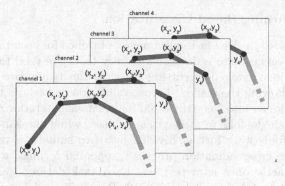

Fig. 2. Example of encoding for the trait generation in MC image.

(frames) of the same handwriting sample, both in air and on paper, into a single image file. Each frame is a grayscale representation of the traits where the color channel encodes a particular dynamic feature. More specifically, considering the points (x_i, y_i) as vertices of the polygonal that approximates the original curve, pixel values in each frame are assigned according to the following criteria (see Fig. 2):

- The first frame implements the acceleration feature: the acceleration of the $i - th$ trait is defined as the derivative of v_i;
- The second frame implements the jerk feature: the jerk of the $i - th$ trait is defined as the second derivative of v_i;
- The third frame implements the velocity feature: the velocity of the $i - th$ trait is computed as the ratio between the length of the $i - th$ trait and interval time of $5\,ms$ corresponding to the frequency of acquisition of the tablet;
- The fourth frame implements the pressure feature at point (x_i, y_i) and it is assumed constant along the $i - th$ trait.

The values of the features (a_i, v_i, z_i, j_i) have been normalized into the range $[0, 255]$ in order to match the standard 0–255 color scale, by considering the minimum and the maximum value on the entire training set for these four quantities.

We used three CNN models that accept input images that are automatically resized to 256×256 for VGG19, to 224×224 for ResNet50 and to 299×299 for InceptionV3. Taking into account these constraints for images size, the original **x, y** coordinates have been resized into the range $[0, 299]$ for each image, in order to provide ex-ante images of suitable size and minimize the loss of information related to possible zoom in/out.

4.2   Deep Learning Based Classification

The aforementioned In-Air_On-Paper images, obtained for each task and for each subject, were arranged into a dataset for every task. We used the 5-Fold Cross validation technique to avoid overfitting and statistically improve the accuracy of the network. Following the standard cross-validation procedure, the dataset was randomly partitioned into 5 equally sized folds. At each iteration, all the samples belonging to a single fold were used as test set, while the samples belonging to the other 4 folds were further divided into two subsets: a validation and a training set. The cross-validation process is repeated 5 times, with each of the 5 folds used exactly once as a test set. Each task dataset was then used to feed three different CNN models: VGG19, ResNet50 and InceptionV3. All of them were pretrained on the public dataset ImageNet to automatically extract features. Those models shared a unique classifier composed of two fully connected layers, properly arranged for our purposes, i.e. classifying two classes (HC or PT) instead of the thousands as is the case of the ImageNet dataset. The classification step is iterated over the 34 tasks so, at the end of this process, every subject will have 34 predictions and their corresponding confidence degrees.

4.3   GA for Task Selection

We considered two approaches using the data obtained as the deep learning classification. First we use a vector of 34 predictions as feature vector of each subject (normal in the following), then we consider the predictions weighted with their corresponding confidence degrees (weighted). To optimize the performance of the system in predicting the cognitive state of the involved subjects, we used a GA to select the subset of 34 tasks for the two adopted approaches, that allows the system to achieve the best prediction performance. As mentioned in the introduction, we used a GA because this algorithm is well-known for its global search ability and also because it provides a natural and straightforward representation of item (tasks in our case) subsets: the value 1 or 0 of the chromosome ith element indicates whether the ith item (task) is included or not in the subset represented by the chromosome. Given the i-th individual to be evaluated, representing the task subset s_i, its fitness was computed by considering only the tasks included in s_i when computing the majority vote rule.

The GA was implemented by using a generational evolutionary algorithm which starts by generating a population of P individuals, randomly generated. Afterwards, the fitness of the generated individuals is evaluated by computing the prediction accuracy on T_r. After this preliminary evaluation phase, a new population is generated by selecting $P/2$ pairs of individuals using the tournament selection method, of size t. The one point crossover operator is then applied to each of the selected pairs, according to a given probability factor p_c. Afterwards, the mutation operator is applied with a probability p_m. Finally, these individuals are added to the new population. The process just described is repeated for N_g generations. Further details can be found in [9].

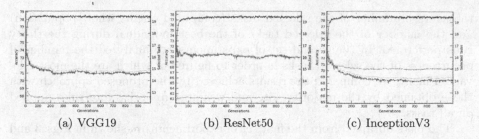

(a) VGG19   (b) ResNet50   (c) InceptionV3

Fig. 3. Evolution of accuracy and average number of selected tasks for every model of CNN, considering weighted predictions.

5   Experimental Results

We tested our approach by training three well-known and widely-used CNNs: VGG19, ResNet50 and InceptionV3. According to the procedure detailed in Subsect. 4.2, we built up a dataset of features for every approach (normal or weighted), where each sample contains the responses provided by the given classifier on the 34 tasks. Each dataset was split into two statistically independent parts: a training set T_r, made of the 80% of the available samples, and a test set T_s, made of the remaining samples. T_r was used to evaluate the individuals' fitness, whereas T_s was used to assess the performance of the best individual on unseen data. For each dataset, we performed thirty runs and at the end of each run, the task subset encoded by the individual with the best fitness was stored as the solution provided by that run. The results reported in the following were computed by averaging those obtained by the thirty best individuals stored. As for the parameters of the GA, we performed some preliminary trials to set them. These parameters were used for all the experiments described below and are reported in Table 1.

We performed three sets of experiments. In the first set, we analyzed the generalization ability of the GA-based system as well as its capability of reducing the number of tasks needed to correctly predict cognitive impairments. To this aim, we plotted the training and test accuracy of the best individual as well as

Table 1. The values of the parameters used in the experiments.

Parameter	Symbol	value
Population size	P	100
Crossover probability	p_c	0.6
Tournament size	t	5
Elitism	e	2
Mutation probability	p_m	0.03
Number of Generations	N_g	1000

the average number of the selected tasks (computed on the whole population) and the number of the selected tasks of the best individual during the thirty runs performed. In the second set of experiments, we analyzed the number of occurrences of the selected tasks in order to figure out which are the most relevant. Finally, we compared the results achieved by the proposed approach with those obtained by the majority-vote and weighted majority-vote rules applied on all tasks.

The plots obtained from the first set of experiments are shown in Figs. 3 and 4. They show the evolution of: (i) the average training (blue) and test (red) accuracy of the best individual; (ii) the average number of selected tasks for the best individual (green) and the whole population (yellow), computed by averaging the values of the thirty run performed. Looking at those plots, it is possible to assess the negative effects of the undesired phenomenon of overfitting. The train and test accuracies, in fact, show similiar trend, as the number of iterations increases, the fitness function increases, but there is a gap between the two accuracies. The performance of the system is better when dealing with training samples than with unseen samples coming from the test set. This means that the proposed system hasn't a great generalization ability, so this surely is an aspect that should be improved. This behaviour is common for every network (VGG19, ResNet50 and InceptionV3) and every approach (Normal and Weighted). The best results about the accuracy trend is obtained considering the weighted predictions coming from InceptionV3, in fact in this case the gap between the train and test accuracy is smaller, also the test accuracy reaches the 71%.

On the other hand, by analysing the other trends shown in the plots, it is easy to assess a good capability of reducing the number of tasks needed for the prediction of cognitive impairments. In fact the number of selected tasks, both average and best, decreases as the number of generations increases. At the end of the run, about less than half of the tasks were selected. Also in this case the best result is obtained with the weighted predictions, coming from InceptionV3, where the lower number of selected tasks is 14. Moreover, it is worth noting that the number of selected tasks of the best individual is always less than the average one. This seems to confirm our assumption that a better prediction performance can be achieved by suitably selecting a subset of the whole set of available tasks.

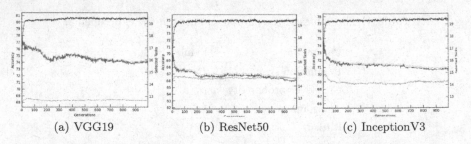

(a) VGG19   (b) ResNet50   (c) InceptionV3

Fig. 4. Evolution of accuracy and average number of selected tasks for every model of CNN, considering non-weighted predictions. (Color figure online)

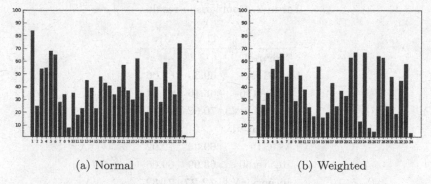

(a) Normal   (b) Weighted

Fig. 5. Comparison of the number of occurrences of the selected tasks for the three CNN between the normal and the weighted approach.

The histograms of the second set of experiments are shown in Fig. 5. They show the number of occurrences of the selected tasks, computed on the thirty runs performed. There is a bar for everyone of the thirty four tasks and its value represents the number of times that task has been selected considering all the CNNs together. The first histogram was obtained using normal predictions, whereas the second comes from weighted predictions. As expected the two histograms show similar trend. It is possible to notice that some tasks are more selected then others, in particular, 1 (signature), 5 (circles), 6 (copy of letters), 22 (copy of phone number), 24 (clock drawing test) and 33 (copy of word). Most of these are graphic or copy tasks, so this research allowed us to highlight the effectiveness of our protocol. The purpose of the graphic tasks is to test the subjects' ability in writing elementary traits, joining some points and drawing figures, while the copy tasks have the objective to test the subjects' ability in repeating complex graphic gesture or motor planning.

In order to test the effectiveness of our system, for the third experiment we compared its results with those achieved by the majority-vote and the weighted majority-vote rules, which are well-known and widely used strategies for combining responses provided by several sources. Those rules combine the predictions made on the tasks selected by the GA, considering normal or weighted predictions. Given a set of responses to be combined, in our case the predictions provided by different tasks, for the single subject, assigns an unknown sample (a subject in our case) to the class (PT or HC) that has the highest occurrence among the predictions coming from the subset of tasks. As for the second rule, it takes into account the prediction performance of the single classifiers by multiplying each response with its confidence degree. Comparison results are shown in Table 2. From the Table 2, it can be seen that in most cases, the majority vote

Table 2. Comparison results.

	GA	Mjv
Normal		
VGG19	69.30	69.06
ResNet50	68.16	65.74
InceptionV3	70.62	70.16
Weighted		
VGG19	69.04	72.92
ResNet50	63.99	69.06
InceptionV3	**72.27**	71.82

rule applied on the tasks selected by the GA, shows a higher accuracy than the majority vote rule applied on the whole number of tasks (34), confirming the effectiveness of our system. The combination rule applied on the subset of tasks selected by the GA, considering weighted prediction coming from InceptionV3 reached the best accuracy score of 72.27%, consisting in the best overall prediction performance. This experiment supports the thesis that a lower number of tasks, of course depending on the chosen tasks, can be enough to distinguish cognitive impairment affected subjects from healthy controls.

6   Conclusions

Cognitive impairments are one the first signs of neurodegenerative diseases, whose incidence is expected to increase in the near feature. Therefore, the improvement of the currently available tools to diagnose these diseases is becoming crucial. Since handwriting is one of the human skills affected by cognitive impairments, recently researches focused on the analysis of handwriting for an early detection of cognitive impairments, achieving promising results.

In this paper, we presented a GA-based approach for the improvement of the performance of a system for the detection of cognitively impaired people. The proposed system records and analyzes the handwriting of people performing some simple tasks. In particular, the GA selects the subset of tasks that allow the maximization of the prediction performance.

In the experiments performed we tested the generalization ability of the system as well as its capability of reducing the number of tasks needed to correctly predict cognitive impairments. Moreover, we also tried to figure out which are the more relevant tasks, i.e. those most frequently selected by the GA in the performed runs. Finally, we compared our results with those of the majority-vote and the weighted majority rules. The experimental results showed that the proposed system has a good generalization ability, by selecting only half of the available tasks. As concerns the relevance of the single tasks, the results proved that copy and graphic tasks are more selected than others. Finally, also the comparison results, coming from the application of the majority vote rule, confirmed the effectiveness of our system.

This research will be extended taking into account further developments. First, we will add new tasks, belonging to the same kind of the most selected ones, into the experimental protocol. In this case the less selected tasks will be substituted. We will also implement some combining strategies [11,13] to improve the performance of our system. Finally, an important improvement can come from different sources of data, e.g. predictions coming from RGB images, which can encode the dynamic features of in air, on paper or both traits.

References

1. Bakurov, I., Castelli, M., Vanneschi, L., Freitas, M.J.: Supporting medical decisions for treating rare diseases through genetic programming. In: Kaufmann, P., Castillo, P.A. (eds.) EvoApplications 2019. LNCS, vol. 11454, pp. 187–203. Springer, Cham (2019). https://doi.org/10.1007/978-3-030-16692-2_13
2. Bojarczuk, C.C., Lopes, H.S., Freitas, A.A., Michalkiewicz, E.L.: A constrained-syntax genetic programming system for discovering classification rules: application to medical data sets. Artif. Intell. Med. **30**(1), 27–48 (2004)
3. Carmona-Duarte, C., Ferrer, M.A., Parziale, A., Marcelli, A.: Temporal evolution in synthetic handwriting. Pattern Recogn. 68(Supplement C), 233–244 (2017)
4. Castelli, M., Vanneschi, L., Manzoni, L., Popovič, A.: Semantic genetic programming for fast and accurate data knowledge discovery. Swarm Evol. Comput. **26**, 1–7 (2016)
5. Cilia, N.D., De Stefano, C., Fontanella, F., Scotto di Freca, A.: Variable-length representation for EC-based feature selection in high-dimensional data. In: Kaufmann, P., Castillo, P.A. (eds.) EvoApplications 2019. LNCS, vol. 11454, pp. 325–340. Springer, Cham (2019). https://doi.org/10.1007/978-3-030-16692-2_22
6. Cilia, N., De Stefano, C., Fontanella, F., Scotto Di Freca, A.: An experimental protocol to support cognitive impairment diagnosis by using handwriting analysis. In: Procedia Computer Science, Proceeding of The 8th International Conference on Current and Future Trends of Information and Communication Technologies in Healthcare (ICTH), pp. 1–9. Elsevier (2019)
7. Cilia, N.D., De Stefano, C., Fontanella, F., Molinara, M., Scotto Di Freca, A.: Handwriting analysis to support Alzheimer's disease diagnosis: a preliminary study. In: Vento, M., Percannella, G. (eds.) CAIP 2019. LNCS, vol. 11679, pp. 143–151. Springer, Cham (2019). https://doi.org/10.1007/978-3-030-29891-3_13
8. Cilia, N.D., De Stefano, C., Fontanella, F., Molinara, M., Scotto Di Freca, A.: Using handwriting features to characterize cognitive impairment. In: Ricci, E., Rota Bulò, S., Snoek, C., Lanz, O., Messelodi, S., Sebe, N. (eds.) ICIAP 2019. LNCS, vol. 11752, pp. 683–693. Springer, Cham (2019). https://doi.org/10.1007/978-3-030-30645-8_62
9. Cilia, N.D., De Stefano, C., Fontanella, F., Di Freca, A.S.: Using genetic algorithms for the prediction of cognitive impairments. In: Castillo, P.A., Jiménez Laredo, J.L., Fernández de Vega, F. (eds.) EvoApplications 2020. LNCS, vol. 12104, pp. 479–493. Springer, Cham (2020). https://doi.org/10.1007/978-3-030-43722-0_31
10. De Stefano, C., Fontanella, F., Folino, G., di Freca, A.S.: A Bayesian approach for combining ensembles of GP classifiers. In: Sansone, C., Kittler, J., Roli, F. (eds.) MCS 2011. LNCS, vol. 6713, pp. 26–35. Springer, Heidelberg (2011). https://doi.org/10.1007/978-3-642-21557-5_5

11. De Stefano, C., Fontanella, F., Folino, G., Scotto Di Freca, A.: A Bayesian approach for combining ensembles of GP classifiers. Lecture Notes in Computer Science. Multiple Classifier Systems. MCS 2011 6713, 26–35 (2011)

12. De Stefano, C., Fontanella, F., Marrocco, C.: A GA-based feature selection algorithm for remote sensing images. In: Giacobini, M., et al. (eds.) EvoWorkshops 2008. LNCS, vol. 4974, pp. 285–294. Springer, Heidelberg (2008). https://doi.org/10.1007/978-3-540-78761-7_29

13. De Stefano, C., Fontanella, F., Marrocco, C., Scotto Di Freca, A.: A hybrid evolutionary algorithm for Bayesian networks learning: an application to classifier combination. Lecture Notes in Computer Science. Applications of Evolutionary Computation. Evo Applications 2010. 6024, 221–230 (2010)

14. De Stefano, C., Fontanella, F., Impedovo, D., Pirlo, G., Scotto di Freca, A.: Handwriting analysis to support neurodegenerative diseases diagnosis. Rev. Pattern Recogn. Lett. **121**, 37–45 (2019)

15. Ghaheri, A., Shoar, S., Naderan, M., Hoseini, S.S.: The applications of genetic algorithms in medicine. Oman Med. J. **30**(6), 406–416 (2015)

16. Impedovo, D., Pirlo, G.: Dynamic handwriting analysis for the assessment of neurodegenerative diseases: a pattern recognition perspective. IEEE Rev. Biomed. Eng., pp. 1–13 (2018)

17. Impedovo, D., Pirlo, G., Vessio, G., Angelillo, M.T.: A handwriting-based protocol for assessing neurodegenerative dementia. Cognitive Comput. **11**(4), 576–586 (2019)

18. Johnson, P., et al.: Genetic algorithm with logistic regression for prediction of progression to Alzheimer's disease. BMC Bioinformatics 15(S11) (2014)

19. Marcelli, A., Parziale, A., Santoro, A.: Modeling handwriting style: a preliminary investigation. In: 2012 International Conference on Frontiers in Handwriting Recognition, pp. 411–416 (Sept 2012)

20. Marcelli, A., Parziale, A., Santoro, A.: Modelling visual appearance of handwriting. In: Petrosino, A. (ed.) ICIAP 2013. LNCS, vol. 8157, pp. 673–682. Springer, Heidelberg (2013). https://doi.org/10.1007/978-3-642-41184-7_68

21. Marcelli, A., Parziale, A., Senatore, R.: Some observations on handwriting from a motor learning perspective. In: 2nd International Workshop on Automated Forensic Handwriting Analysis (2013)

22. Neils-Strunjas, J., Groves-Wright, K., Mashima, P., Harnish, S.: Dysgraphia in Alzheimer's disease: a review for clinical and research purposes. J. Speech Lang. Hear. Res. **49**(6), 1313–30 (2006)

23. Senatore, R., Della Cioppa, A., Marcelli, A.: Automatic diagnosis of neurodegenerative diseases: an evolutionary approach for facing the interpretability problem. Information **10**(1), 30 (2019)

24. Senatore, R., Della Cioppa, A., Marcelli, A.: Automatic diagnosis of Parkinson disease through handwriting analysis: a cartesian genetic programming approach. In: 2019 IEEE 32nd International Symposium on Computer-Based Medical Systems (CBMS), pp. 312–317 (2019)

25. Tseng, M.H., Cermak, S.A.: The influence of ergonomic factors and perceptual-motor abilities on handwriting performance. Am. J. Occup. Therapy **47**(10), 919–926 (1993)

26. Valenzuela, O., Jiang, X., Carrillo, A., Rojas, I.: Multi-objective genetic algorithms to find most relevant volumes of the brain related to Alzheimer's disease and mild cognitive impairment. International Journal of Neural Systems 28(09) (2018)

Autonomous Inspections of Power Towers with an UAV

Guillaume Maître[1]([✉])(iD), Dimitri Martinot[2], Dimitri Fontaine[2],
and Elio Tuci[1](iD)

[1] University of Namur, Namur, Belgium
{guillaume.maitre,elio.tuci}@unamur.be
[2] Qualitics SPRL, Ans, Belgium
{dimitri.martinot,dimitri.fontaine}@qualitics.eu
https://www.unamur.be/info/

Abstract. Electrical grid maintenance is a complex and time consuming task. In this study, we test a way to perform electrical grid tower inspection by using camera images taken from an autonomous UAV. The images are segmented with a type of convolutional neural network called U-Net [9]. The results of the segmentation process is used to generate the movements of the UAV around the tower. The training of a U-Net model requires a large amount of labelled images. In order to reduce the time and financial costs of the generation of a large data-set of labelled images of physical towers, we develop a physics-based simulation environment that models the UAV dynamics and graphically reproduces electric towers in multiples environmental conditions. We extract labelled images for U-Net models training from the simulator. We perform multiple training, test conditions with different amount of natural world and simulated images and we evaluate which training condition generates the most effective U-Net model for the natural world image segmentation task. The contribution of the study is to detail the characteristics of the training condition that allows to maximise the U-Net performances with the minimum amount of physical world images in the training set. With the best performing U-Net, we create a post-processing analysis on the result of the segmentation to extract the required pieces of information to properly move the UAV.

Keywords: Computer vision · Convolutional neural network · Autonomous system

1 Introduction

UAVs (Unmanned Aerial Vehicle) are particularly suitable to automate the exploration and monitoring of remote areas such as glaciers, and volcanos, but also the inspection of infrastructures which would otherwise require complex

G. Maître—Thanks the Walloon Region for the financial support "Doctorat-en-Entreprise" fellowship.

J. J. Schneider et al. (Eds.): WIVACE 2021, CCIS 1722, pp. 151–162, 2022.
https://doi.org/10.1007/978-3-031-23929-8_15

and/or costly operations if carried out without UAVs [2,6]. In this latter case, the inspection process is generally performed with cameras or other sensors mounted directly on the UAV, which flies around the target in order to optimise the data collection process relative to its status [8]. The objective of our work is to develop new methodologies to improve the efficiency as well as the autonomy of UAVs during high-precision inspection tasks. In particular, we describe a set of experiments aimed to automate the inspection of electricity grid pylons by using autonomous and visually-guided UAVs. The vehicle gets close to and flies all around a target pylon while collecting images which are, in the first instance, used by the UAVs itself to navigate and to make an exhaustive exploration of the target while avoiding collisions with any object. The images are also collected and processed offline to create SVG images of the precise pylon shape, and to detect damaged areas or any element requiring further attention.

Currently, electric pylon inspection is a critical and costly task due to the time and effort required. If we exclude the use of UAVs, the inspection process can be performed in one of the following ways:

- By helicopter: This is an extremely expensive method. It requires expensive specific material such as a camera with high distance zoom and different experts to pilot, to take photos and to analyse the collected data. This method does not provide a bottom-up point of view of the electric pylon since the helicopter can not fly underneath the cables.
- Unaided-eye visual inspection from ground: This method only provides a mild inspection and only from a bottom-up perspective. Nevertheless, it is the cheapest and also the fastest method.
- By climbing the tower: This is the more informative method, which gives a clear insight of the state of the electric pylon. Unfortunately, it requires shutting down the electric line completely. Due to this requirement, this method is the rarest one employed. This method is also the most dangerous, since some electric pylons might be too rusty and metal bars could collapse under the weight of the climber.

The automation of the inspection process of electric pylon by UAVs presents several clear advantages over the above mentioned alternative methods. With autonomous UAVs, the human factor is completely removed from the inspection process by significantly improving the safety and comfort of the personnel, without the need to shut-down the electric line. Plus, images or any other types of data relative to the pylons can be taken from perspectives generally inaccessible to non-UAV-based inspection systems. Also, the automation enables a consistency and precision that a human remote pilot can hardly achieve, thus improving the quality of the audit and the accuracy of the diagnosis.

In this paper, we investigate the issue of how to train a convolution neural network (CNN) to identify the structure of pylons from images taken by a camera mounted on the UAV. The image segmentation process has to be informative enough to allow the UAV to autonomously move around the pylons without crashing into their structure or into the electric cables supported by the electric pylon. Our CNN-based controller is an U-Net model [9], a CNN model that can perform real time segmentation without requesting much data for training. The CNN guides the UAV in the task of detecting the electric pylon and in exhaustively exploring all its parts.

U-Net is made of two parts: a contraction path to extract a maximum of features and contexts out of an image; and a symmetric expanding path to extract a pixelwise precision mask of the object in the image. A mask is an image in which the intensity of pixels corresponding to a target object is set to non-zero values and all the other pixels are set to zero. The main advantage of U-Net is the capacity to work with few available images in a training data-set. A series of comparative tests detailed in [9] shows that U-Net has been able to outperform other types of convolutionnal neural networks in the 2015 IEEE-ISBI cell tracking challenge. The model proved to be able to perform precise cell detection in less than a second for 512*512 size images on 2015 GPUs (NVIDIA GTX Titan 6 Gb) [9].

Compared to other object detection models, such as Fast R-CNN [5], which create bounding boxes around the object to be detected, U-Net proposes a pixel-wise detection which precisely identifies the object within an image. The disadvantage of U-Net is that it does not offer the possibility for multi-instance object segmentation. This means that it is not possible to extract different bodies of the same type in an image or similar objects that are next to each other. For this kind of tasks, it is suggested to use other convolutional neural network models like Fast R-CNN. However, given that electric pylons are usually represented as a single instance object in images, U-Net is a suitable model for their identification through detection in camera images. Thanks to the segmentation process and analysis with few masks, it is possible to retrieve important information not only concerning the conditions of the electric pylon but it is also possible to easily determine the position of the camera (and consequently of the drone on which the camera is mounted) with respect to the electric pylon.

In this paper, we look at a way to use U-Net to perform a precise enough mask of an electric pylon in an image. The complexity of this task is determined by the shape of the electric pylon which, due to the void space between the electric pylon's structures, makes it hard to be properly detected. Moreover, the creation of large data-sets of natural world images of electric pylons, which are generally required for the training of convolution neural networks, is a costly and difficult task due to the relatively large variability in the shape of electric pylon. With respect to this challenges, U-net offers several advantages compared to other models since it can successfully be trained to perform object segmentation tasks without having to rely on large data-sets. Moreover, U-Net can perform the image segmentation tasks quite fast without requiring particularly powerful computational resources.

The contribution of this study is to show that U-Net can be effectively trained for this inspection task with hybrid sets of images in which the large majority is generated with a simulation environment that models the main features of the target scenarios. To generate the simulated images, we make use of AirSim [11], a simulation software developed by Microsoft powered by Unreal Engine. In particular, we show that the training performed with sets of simulated images is as effective in the segmentation as with sets of images from the real world. In the following sections, we first describe the simulation environment used to generate the simulated images and the experimental design used to test the hypothesis that hybrid sets of simulation-generated and physical world images are as effective for the training as sets of only physical world images (see Sect. 2). In Sect. 3 we illustrate and discuss the results of our analysis. In Sect. 4, we draw our conclusions.

2   Methods

In this section, we illustrate the methodological aspects of our study. First of all, we introduce the simulator used to create the simulated scenarios. Then, we describe the characteristics of the UAV employed. After that, we explain the architecture of the U-Net model and the analysis performed on the camera images to generate the UAV movement.

2.1   The Simulator and the UAV's Model

(a)   (b)

Fig. 1. Images from the simulator rendering: (a) an urban scenario; (b) a rural scenario.

The simulation environment is created using the software AirSim [11], made of Unreal Engine physics engine and high quality graphics for rendering, with which we model the dynamics of the flying vehicle, the pylon of the electricity grid and details of the surrounding area. For this study, we have modelled a single type of pylon, located in multiple types of background, one urban, one forest, two rural, and one mountain scenario (see Fig. 1). In each scenario, there are three electric pylons, spaced 400 m apart. In one of the rural scenario, there is also a windmill with moving blades which includes dynamic elements in the scene. In every type of scenario, the horizon is represented by distant mountains which are drawn in a very ragged way to avoid the network to exploit the horizontal skyline for its movements. Each scenario is also replicated in different weather and luminosity conditions (e.g., with sun, rain, snow, fog, etc.). The variability in the simulated scenarios in which the electric pylons are placed is needed to make sure that the convolutional neural network can identify the structure of the tower regardless of the characteristics of the background and meteorological conditions.

As far as the structure of the UAV is concerned, AirSim has already a built-up UAV model (quadcopter) and offers the possibility to add our own models. The built-in model has the possibility to be customised in terms of sensorial capabilities. In this study we use the built-in quadcopter with a front facing camera that generate images with a resolution of 5280*3956 pixels; the FOV is 72°; the focal distance is 200.0 and the focal region is 200; no specific filter is applied to the image. The convolutional neural network performs the segmentation of the images generated by the camera in real time in order to identify the pixel-wise position of the electric pylon in the image. A hand-designed control system generates high level instructions from the results of the segmentation process. Both the segmentation process and the rules to generate the UAV movements are described below.

The PixHawk PX4 fight controller takes care of regulating the speed of the rotors in order to execute the high level motor commands by taking into account the physical features of the UAV. Thanks to the PX4 flight controller, the behaviour of the simulated UAV can be easily replicated in a physical UAV. The UAV is also equipped with a ground facing radar to measure the distance from ground or any object placed underneath its body. The information generated by the radar is used to manoeuvre the UAV safely (without collisions) while moving around the pylon. The UAV is also equipped with a seven points front-facing LiDAR to precisely measure the distance between the UAV and any object in the space in front of it. The LiDAR makes it possible to keep the UAV always at more than 4 m away from any element of the electric pylon.

For what concerns the movements, the UAV moves by using PIDs for pitch and roll. A constant value is used for height adjustment.

The movement of the UAV around the electric pylon is illustrated in Fig. 2. The UAV stabilises and adjusts to be centred at the same height as the upper part of the pylon (see position 1 in Fig. 2). While in position 1, the camera starts taking snapshots which are processed by the convolutional neural network to detect the position of the pylon. The results of the image-segmentation process is used by the control system to move the UAV downward while remaining at a constant safe distance from the pylon. When the down facing Radar get close enough to the ground or an obstacle and the front facing LiDAR detects no object in front of the UAV (see position 2 in Fig. 2), the downward movement is stopped. At this point the UAV generate a parabolic movement to reach the other side of the pylon moving below the electric cables (see position 3 in Fig. 2). Once of the other side, the UAV starts an ascending movement, always using the results of the image-segmentation process to remain in the proximity of the pylon. The ascending movement terminates when the UAV reaches the top of the electric pylon (see position 4 in Fig. 2). From position 4 the UAV returns to position 1 following the same trajectory. This exploration pattern makes it

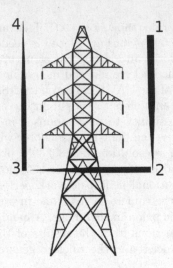

Fig. 2. Schematic view of the electricity pylon with markers to represent the UAV's movements during exploration.

possible to have images of the two side of the electric pylon taken from both the ascending and descending movement of the UAV.

As the UAV is following this path, the UAV takes an image (resized to 320×320 pixels) once every 130 milliseconds (see Fig. 3a). Each images is segmented by the U-net to generate a pixelwise segmented images in which pixels belonging to the pylon are in white and the background is black (see Fig. 3b). Onto these images we apply the Canny Edge detection algorithm [3] to extract the border of the pylon. Then, with a Probabilistic Hough Line Transform [7], we extract the lines of the border of the electric pylon (see Fig. 3c).

After this processing, we can extract the information necessary to move the UAV in order to have the pylon at the centre of the image. To compute the UAV's distance from the electric pylon from the image we compute width x of the tower (in pixels) and we compare x with the expected width of the electric pylon dist for the tower in 4 m distance from the UAV, where $dist$ is computed as

$$dist = \left(\left(\frac{y}{\tan(FOV \times z)} * 3,2 \right) * 100 \right) \tag{1}$$

where y is the width of the cage of the pylon in centimetres and $z = 4$ represent the security distance between the UAV and the pylon. If x is bigger than $dist$, the UAv is too close to the tower. If instead x is smaller than $dist$, the UAV is more than 4 m away from the tower.

2.2   The U-Net and Our Experimental Design

In order to train the U-Net to identify the tower in the images we have generated 1200 images, out of which 400 images from the natural world and 800 generated

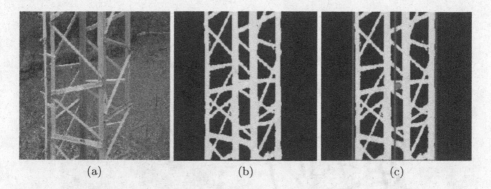

(a)   (b)   (c)

Fig. 3. (a) An image taken by the UAV's camera; (b) The image in (a) after segmentation by the U-Net; (c) The image from (b) after post-processing with Canny-type edge detection and application of the Hough-Transform.

from the simulator. This data set benefits from data augmentation (e.g., flipping, rotation and removing one of the colour channels from the RGB spectrum, etc) for the original images. The natural world images have been manually labelled in order to identify pixels belonging to the tower and those belonging to the background. This labelling process is a long and relatively expensive process that is however necessary for the training. The images from the simulator are labelled automatically.

Our experimental design features four training conditions and three test conditions described in Table 1. The training conditions differ in the number of natural world and simulated images used for training. Note that in the training condition A we do not use any natural world images. Condition D is the one with the higher number of natural world images. We assume that the highest the number of natural world images in the training set, the better the network performs in the segmentation task. The objective of this study is to verify whether and the extend to which natural world images can be replaced with simulated images without significant loss of performance. For this reason, the performance of the U-Net trained in condition A, B, and C will be compared with the

Table 1. Table summarising the experimental design with the four training conditions and the three test conditions

Data-set name	% simulated images	% physical images
Training A	100	0
Training B	80	20
Training C	90	10
Training D	50	50
Test I	50	50
Test II	100	0
Test III	0	100

(a)   (b)

Fig. 4. (a) Natural world image used for trainig/test and (b) simulated image.

performance of the U-net trained in condition D. Note that, we do not have a training condition determined by all natural world images simply because the number of these images at our disposal at the time of this study was not sufficient to train the network.

The training sets were also divided with a k-fold into a 80–20 percent training and validation sets by using a k-fold data loader during the training of the models. The structure of the U-Net is detailed in Fig. 5. The U-Net takes as input RGB images of 320*320 pixels and generate a vector of 320*320 real-valued numbers between 0 and 1. These values represent the probability that a pixel belong to the tower. The U-net is modelled using the programming framework TensorFlow [1]. We have tested different parametrisations of the training process, by varying the number of epochs, the function optimiser, the nature of the loss function, the size of the data-set, the training stop-criteria (see Table 2 for details).

3   Results

From all the possible values represented in Table 2, we tried every combination of parameters. Each training condition illustrated in Table 1 has been replicated for

Table 2. The training parameters

Parameter	Tested values
Epoch	25, 50, 75
Optimiser	AdaMax, Adam, Nadam
Loss function	Mean square error, Absolute square error
Images in the training data set	400, 800, 1600
Early stopper	None, 2,3,4

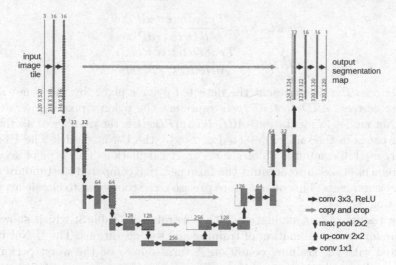

Fig. 5. The structure of the U-Net model used in the study

each possible combination of the parameters listed in Table 2. In this section, we only discuss the results generated from the best set of parameters with 50 epochs with an early stopper of 2; a total of 800 images in the training set, Adam and the Mean Square Error as optimiser as well as a loss function in the F1-score. Recall that the objective of our study is to train the U-Net model to be able to extract the information required to move the UAV around an electric pylon by using primarily computer vision. To do so, the U-Net model has to be able to properly segment the images to precisely determine where the pylon is and to perform the right movements for the audit. To find out the best performing U-Net model, we experimentally vary the proportion of physical world and simulated images in the training set.

The performances of the U-Net model are evaluated with the F1-score. This score is computed in the following:

$$F1 = \frac{2}{recall^{-1} + precision^{-1}} \tag{2}$$

Table 3. Table showing the F1-score for each combination of training and test condition.

Data-set name	Test I	Test II	Test III
Training A	0.485666	0.74248765	0.25558049
Training B	0.807414	0.86110346	0.81913397
Training C	0.767606	0.80255722	0.79024012
Training D	0.848198	0.82709646	0.89200856

$$Precision = \frac{TrueDetectedPixels}{AllDetectedPixels} \tag{3}$$

$$Recall = \frac{TrueDetectedPixels}{AllRelevantPixels} \tag{4}$$

TrueDetectedPixels represent the detected pylon's pixels by the U-net model that are correct. *AllDetectedPixels* represent the pylon's pixels generated by the U-Net model segmentation. *AllRelevantPixels* is the number of all the real pylon's pixels in the image, detected or not by the U-Net model. The F1-score is wildly used in computer vision for accuracy calculation. This type of accuracy measurements takes into account the false negative compared to standard accuracy measurements. This offer a more precise score compared to classic accuracy calculation.

The results of our simulations are illustrated in Table 3, which shows the F1-score for each combination of training and test conditions. The U-Net model generated with the training condition A turns out to be the worst performing model in all possible test conditions. This relatively negative performance can be accounted for by the reduced variability in the scenarios represented in the training set. Such reduced variability generates what is referred to as an "over-fitting", since the model can not generalise its performance beyond the images presented in the training condition. The training condition B and C generated better performing U-Net models than condition A, with condition B being the best one in all possible test conditions. The lower F1-score of the U-Net from training condition C can also be accounted for by some "over-fitting", caused by the reduced number of physical world images compared to condition B. When comparing training conditions B and D, training condition D outperform B in test condition I by 0.04 and test condition III by 0.07. This shows that more natural world images enable the U-Net model to perform better on natural world images as well as natural world images and simulated images together. It is important that the model can perform well while remaining as inexpensive as possible in terms of cost and time. So, even if the training condition D is performing better, the training condition B is the most valuable for us. Training condition B offers the best cost to performance ratio since it only require 160 natural world images compared to 400 for training condition D. As far the UAV's movement is concerned, with the U-Net model trained with the training condition B we are able to perform all the movement required to perform an audit by following the trajectory illustrated in Fig. 2. The sequence of movements requires between 4 min and 24 s and 6 min 21 s depending on the scenarios and the starting position. The UAV is able to keep its safe distance with the pylon without much difficulties. The difference in time is due to different starting positions which require the UAV to centre itself before starting and the search for obstacles that can activate the position 2 of Fig. 2 earlier.

4   Conclusions

We achieved our main goal during this study which was to train a U-Net model to visually guide an autonomous UAV during an inspection task of an electric grid pylon. The main challenge was to find a dataset which could be processed economically and fast of labelled images to train our U-Net model. To do so, the model was trained with different types of data-sets made of a different number of natural world and simulated images mixed together.

The U-Net model generated with training condition B is performing relatively well with both natural world and simulated scenarios. The mixing of both natural world and simulated images helps the U-Net model to reduce the effect of the Reality Gap [10] for the world perception through the camera. The training condition B and D are doing well enough to perform similarly in the natural world and the simulated world. We conclude that the training condition B is better even with a lower score than training condition D. The cost and the time required to create the training condition B are both factors to prefer such a condition for training. It maintains good overall score while only requiring 0.2% of natural world images.

In terms of the F1-Score, a score superior to 0.8 is considered good enough for a convolutional neural network to generalise the segmentation capabilities beyond the training set. The pylons are a really complex shape with a lot of details and smaller parts. That complex shape makes it difficult to properly segment the pylon especially with the 320*320 pixels image. With a score superior to 0.8 the post-processing has the required information to generate all the requested data for the movement.

In our future work, we plan to test this system on a natural world electric pylon and see how well the UAV is able to perform when facing other elements not taken into account in our images (e.g., wind, sun glare, sensors noise, etc). Those natural world experiments are expected to show the more complex interactions that can change the behaviour of the UAV compared to the simulation. Moreover, since the time where we ran our experiments, we have collected a lot more natural world images. With those additional images, we could test different neural networks that would not properly work with the few images we had during this study. Concerning the inspection of the pylon, we will be looking at a way to perform instance object detection for flaws (e.g., rust, damage, missing component, bird nest, etc). Automated advanced inspections would help to create a clear database of the state of the electric grid pylon for the owner. The final objective is to create a data-driven self-regulating systems where a UAV detects the flaws and multiple other UAVs would be able to react to such detection to perform the required repairs. Another future work would be to look at more complex models to extract depth in the image while still using one camera, (e.g., M4Depth [4]). This way, we could do without the use of LiDAR by combining both the U-Net model and the M4Depth model to extract proper distance information for security and the position of the pylon. A better understanding of the surrounding can also helps the UAV to perform better obstacle avoidance in more complex scenarios.

References

1. Abadi, M., et al.: TensorFlow: large-scale machine learning on heterogeneous systems (2015). https://www.tensorflow.org/, software available from tensorflow.org
2. Albani, D., Manoni, T., Arik, A., Nardi, D., Trianni, V.: Field coverage for weed mapping: toward experiments with a UAV swarm. In: Compagnoni, A., Casey, W., Cai, Y., Mishra, B. (eds.) BICT 2019. LNICST, vol. 289, pp. 132–146. Springer, Cham (2019). https://doi.org/10.1007/978-3-030-24202-2_10
3. Canny, J.: A computational approach to edge detection. IEEE Trans. Pattern Anal. Mach. Intell. 8(6), 679–698 (1986). https://doi.org/10.1109/TPAMI.1986.4767851
4. Fonder, M., Ernst, D., Droogenbroeck, M.V.: M4Depth: a motion-based approach for monocular depth estimation on video sequences (May 2021)
5. Girshick, R.: Fast R-CNN. In: Proceedings of the IEEE International Conference on Computer Vision (ICCV) (December 2015)
6. Hallermann, N., Morgenthal, G.: Unmanned aerial vehicles (UAV) for the assessment of existing structures, vol. 101 (09 2013). https://doi.org/10.2749/222137813808627172
7. Kiryati, N., Eldar, Y., Bruckstein, A.: A probabilistic Hough transform. Pattern Recogn. 24(4), 303–316 (1991). https://doi.org/10.1016/0031-3203(91)90073-E, http://www.sciencedirect.com/science/article/pii/003132039190073E
8. Máthé, K., Buşoniu, L.: Vision and control for UAVs: a survey of general methods and of inexpensive platforms for infrastructure inspection. Sensors 15(7), 14887–14916 (2015). https://doi.org/10.3390/s150714887, https://www.mdpi.com/1424-8220/15/7/14887
9. Ronneberger, O., Fischer, P., Brox, T.: U-Net: Convolutional networks for biomedical image segmentation (2015)
10. Scheper, K.Y.W., Croon, G.D.: Abstraction, sensory-motor coordination, and the reality gap in evolutionary robotics. Artif. Life 23, 124–141 (2017)
11. Shah, S., Dey, D., Lovett, C., Kapoor, A.: Airsim: high-fidelity visual and physical simulation for autonomous vehicles. CoRR abs/1705.05065 (2017). http://arxiv.org/abs/1705.05065

Self-organizing Maps of Artificial Neural Classifiers - A Brain-Like Pin Factory

Gabriel Vachey and Thomas Ott$^{(\boxtimes)}$ (iD)

Zurich University of Applied Sciences, Institute of Computational Life Sciences,
8820 Wädenswil, Switzerland
ottt@zhaw.ch

Abstract. Most machine learning algorithms are based on the formulation of an optimization problem using a global loss criterion. The essence of this formulation is a top-down engineering thinking that might have some limitations on the way towards a general artificial intelligence. In contrast, self-organizing maps use cooperative and competitive bottom-up rules to generate low-dimensional representations of complex input data. Following similar rules to SOMs, we develop a self-organization approach for a system of classifiers that combines top-down and bottom-up principles in a machine learning system. We believe that such a combination will overcome the limitations with respect to autonomous learning, robustness and self-repair that exist for pure top-down systems. Here we present a preliminary study using simple subsystems with limited learning capacities. As proof of principle, we study a network of simple artificial neural classifiers on the MNIST data set. Each classifier is able to recognize only one single digit. We demonstrate that upon training, the different classifiers are able to specialize their learning for a particular digit and cluster according to the digits. The entire system is capable of recognizing all digits and demonstrates the feasibility of combining bottom-up and top-down principles to solve a more complex task, while exhibiting strong spontaneous organization and robustness.

Keywords: Self-organizing maps · Artificial neural classifiers · Bottom-up principles

1 Introduction

The great advances and successes of machine learning (ML) methods in the last decade, especially in the field of Deep Learning, may hide fundamental peculiarities and possible limitations of these approaches towards more general artificial intelligence. One peculiarity has received surprisingly little attention so far: most ML algorithms are optimized according to a single objective function in a top-down engineering approach. In contrast, there are biologically inspired bottom-up principles of development or learning in systems, such as Hebb's rule [1], which are not necessarily derived from an explicit objective function. Such natural complex systems are associated with spontaneous order, a lack of central

© The Author(s), under exclusive license to Springer Nature Switzerland AG 2022
J. J. Schneider et al. (Eds.): WIVACE 2021, CCIS 1722, pp. 163–171, 2022.
https://doi.org/10.1007/978-3-031-23929-8_16

control, robustness and self-repair [2]. It is these qualities that interest us in this contribution: How can we leverage bottom-up principles for ML?

Nature is a great role model. Almost every biological organism develops from a single pool of elementary cells that, through a process of differentiation, specialize into coherent and functional organs and thus form a viable organism. The principle of spatial self-organization is particularly well represented by the cortical visual field. Indeed, regions within the visual cortex are organized in such a way that their spatial distribution maps the image acquired by the retina [3]. A similar principle is used by Self-Organizing Maps (SOMs) [4], a type of artificial neural networks (ANNs) that are trained to produce low-dimensional representations of the input space. SOMs use competitive and cooperative rules to cluster the input space into discrete categories. However, it is primarily the purpose of visualization that has made SOMs very popular for a wide range of applications (see e.g. [5,6]). The cortical visual field and SOMs are not only examples of the emergence of order from self-organization principles, they also show the characteristics of a 'division of labor' through a process of differentiation. Subsystems, in this case single neurons, specialize to represent a specific input pattern.

In this contribution, we adopt the idea of SOM for more complex systems by replacing the artificial single neurons with artificial neural classifiers. Each of these subsystems may not be complex enough to solve a particular classification task. However, by means of differentiation, the assembly can learn to cope with the task while developing a robust topological order. Unlike for SOMs, this order is not the primary target of our approach and appears as a kind of by-product. We employ the SOM principle to achieve a division of labor among weaker units for machine learning tasks in general. This division of labor is also reminiscent of the idea of the mixture of experts principle used in extensions of SOM such as Generative Topographic Mapping [7]. In contrast to all these approaches, our approach does not primarily aim to preserve the topology of the input space. Rather, this is the consequence of the self-organization, which enables division of labor and guarantees robustness in the event of a subsystem failure.

In the following, we present a proof of principle for the idea, laying the foundation for further research.

2   Self-organizing Maps of Networks: Methods

2.1   General Framework

Starting from the general setting of SOM, we define a map of machine learning algorithms, which in our case are neural classifiers (classifier networks) as sketched in Fig. 1. The classifiers are randomly initialized and are confronted with a classification task for k classes. However, each classifier has only one output and can therefore only learn one class. The classifiers can be trained using backpropagation, minimizing a loss function L. In analogy to SOM, the whole network of networks carries out self-organized learning in three basic steps per learning cycle [8].

1. In the **competitive process**, the winner that minimizes the current sample loss L_j is evaluated:

$$i_{win} = \arg\min_j L_j, \tag{1}$$

where j denotes the jth classifier.

2. In the **cooperative process**, the winning classifier excites similar classifiers, e.g. adjacent ones. The cooperation function $h_{i_{win}}(j)$ describes the learning behavior of classifier j caused by the winner i_{win}.

3. In the **adaptive process**, the classifiers try to reduce their losses L_j. The learning behavior of classifier j depends on the cooperation function $h_{i_{win}}(j)$.

The implementation of this basic scheme with some modifications is described in the next section.

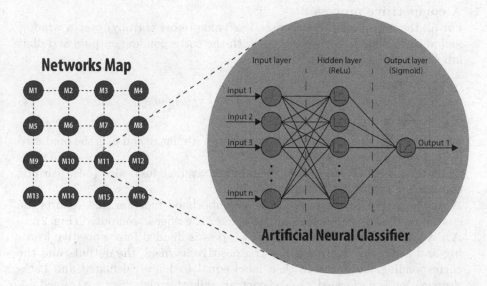

Fig. 1. Networks map of the artificial neural classifiers. ANNs are represented on a map of $n \times m$ classifiers. Each node on the map corresponds to an ANN with a single output neuron. Neurons from the hidden layer use a ReLu activation function while the single output neurons use a sigmod activation function.

2.2   Implementation for Proof of Principle

The different neural classifiers described hereafter have been generated with Tensorflow libraries and Keras API. As proof of principle, we used the MNIST data to train and describe our model. The MNIST data is composed of 28×28 bits images representing digits from 0 to 9. We first generated a simple map of elementary neural networks (represented as a matrix of $n \times m$ Keras sequential classifiers). Each neural network consists of three layers (see Fig. 1). The dimension of the input layer corresponds to the shape of the flattened input image from MNIST data (784 values). The hidden layer consists of a variable number of neurons equipped with the ReLu activation function. Finally, the output layer is one single neuron using the sigmoid activation function. This allows each elementary network to learn only one digit. We built a custom loop in order to dissociate the forward pass and the backpropagation and to adapt the learning process. Since each network is able to learn only one single digit, we used single observation batches as inputs and the labels were one hot encoded. For the backward propagation, we used the Adam optimizer and a learning rate of 0.001.

As for SOMs, we divided the learning into three processes that follow a similar scheme as sketched above:

1. **A competitive process**:
   For all the classifiers j, an average loss value (cross entropy) over a window including the last n learned images with the corresponding outputs and digit labels is calculated (see Fig. 2):

$$L_j = \frac{1}{n} \sum_{i=1}^{n} -[y_i \cdot log(p_i) + (1 - y_i) \cdot log(1 - p_i)] \tag{2}$$

   where y_i is the (one-hot encoded) label of the ith image and p_i is the predicted probability of the label y_i.
   Then the neural classifier with the minimal average loss value (best learning classifier or winner classifier) is identified.
2. **A cooperative process**: A Moore neighborhood function that reports the spatial neighboring classifiers of the winner classifier is computed (Fig. 2).
3. **An adaptive phase**: The adaptive process is divided into a positive learning and a negative learning. For the positive learners, the outputs and the corresponding loss values with a label equal to 1 are calculated and backpropagation is performed. On the contrary, all other classifiers are trained not to recognize the current digit (negative learning). This is done by calculating the loss value with a label equal to zero and performing back-propagation based on these updated loss values.

As validation metric and to follow the learning process across the epochs, we computed a binary accuracy function with the MNIST validation data at the middle and the end of each epoch.

3   Results and Analysis

As proof of principle, we first tested our model with a selection of four digits (0, 1, 2 and 3) and with 25 artificial neural classifiers. The hidden layer of each classifier was composed of 12 neurons. As control, we measured the number of

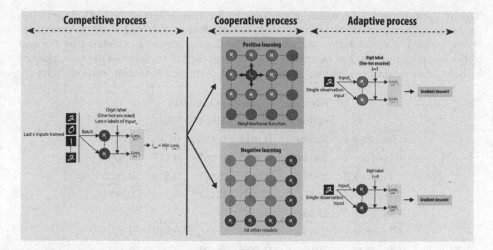

Fig. 2. Training processes The learning rules are based on the principles of SOMs and have been adapted to generate our model. In the competitive phase, we computed for every classifier a loss function over the last 16 input images and the neural classifier with the lowest loss value was identified as the winner classifier. In the cooperative phase, based on the winner classifier, we identified the adjacent neural classifiers with a Moore neighboring function. The selected neural classifiers went through a positive learning while we performed a negative learning for all other neural classifiers. In the adaptive phase, we calculated the loss function for the current image with 1 as label for positive learning and with 0 as label for negative learning. Backpropagation was finally applied to update the weights accordingly.

times each classifier was trained to recognize a digit (positive learning). Table 1 shows that all of the neural classifiers were positively trained with at least one quarter of the training data (i.e., 24754 images). Neural classifiers located in the center part of the map have been trained with a higher number of images while showing less specialization for digits.

The validation accuracy for the different digits shows that most neural classifiers were clearly attributed to a digit while the number of images trained has not a clear impact on accuracy (Table 1). During the learning process, the selection of the best learning classifiers was totally autonomous.

After four epochs, for most neural classifiers, the validation accuracy for a unique digit rapidly increased over time. The different classifiers were thus able to specialize in the recognition of a single digit. The validation accuracy for a unique digit reached values close to 1 (Fig. 2).

Nevertheless, as previously mentioned, some classifiers are not able to distinguish between the digits and their validation accuracy remains below 80% (Fig. 2).

Table 1. Summary of leaning parameters for the different neural classifiers.
As control, we reported the number of images for which each classifier has been positively trained. The validation accuracy for the three digits was calculated at the end of the learning process. For each classifier, the accuracy of the learned digit is depicted in bold. The validation accuracy depicted in red corresponds to non-learned digits.

	Nb. img. trained	Validation accuracy			
		Digit 0	Digit 1	Digit 2	Digit 3
Classifier 1	23689	**0.994**	0.488	0.519	0.519
Classifier 2	29924	**0.994**	0.487	0.517	0.522
Classifier 3	28004	0.591	0.421	0.459	**0.912**
Classifier 4	24472	0.521	0.481	0.514	**0.986**
Classifier 5	18237	0.758	0.721	0.745	0.763
Classifier 6	23934	**0.992**	0.486	0.516	0.521
Classifier 7	30196	**0.996**	0.491	0.52	0.521
Classifier 8	28319	0.54	0.502	0.529	**0.978**
Classifier 9	24711	0.539	0.502	0.529	**0.979**
Classifier 10	18449	0.764	0.727	0.752	0.757
Classifier 11	34552	0.51	**0.976**	0.482	0.485
Classifier 12	39802	0.764	0.727	0.752	0.757
Classifier 13	37333	0.764	0.727	0.752	0.757
Classifier 14	28643	0.764	0.727	0.752	0.757
Classifier 15	23393	0.764	0.727	0.752	0.757
Classifier 16	26916	0.493	**0.996**	0.482	0.485
Classifier 17	27002	0.492	**0.997**	0.481	0.485
Classifier 18	27153	0.764	0.727	0.752	0.757
Classifier 19	23939	0.525	0.489	**0.979**	0.526
Classifier 20	23853	0.535	0.497	**0.974**	0.533
Classifier 21	26671	0.491	**0.996**	0.481	0.484
Classifier 22	26730	0.49	**0.993**	0.481	0.486
Classifier 23	26838	0.764	0.727	0.752	0.757
Classifier 24	23700	0.537	0.501	**0.972**	0.533
Classifier 25	23641	0.527	0.497	**0.975**	0.522

The different neural classifiers have a spatial localization that is the basis for the Moore neighborhood function. In order to show the spatial specialization of the different networks, for each classifier, we calculated the validation accuracy of every digit and reported the digit with the highest accuracy on a map (Fig. 3).

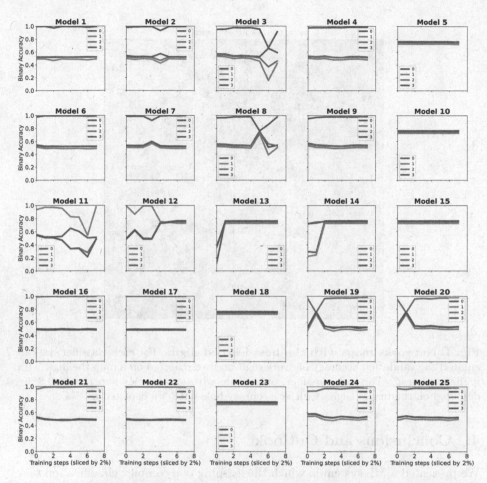

Fig. 3. Validation accuracy of the four digits for the different neural classifiers over training steps. A map of 25 artificial neural classifiers was trained on the MNIST training images (selection of digits 0, 1, 2 and 3). The autonomous specialization of the networks for a particular digit was performed by combining competitive and collaborative rules as described in the implementation section. We measured the validation accuracy for each digits at the middle and the end of each epoch.

Networks with an accuracy below 0.8 were marked with 'Na' as we can consider that below this threshold the digit is not properly learned. Through training, the networks self-organize to form clusters of specialized networks for a particular digit. (Fig. 3). The non-learning classifiers stands in the vicinity between groups of specialized networks (Fig. 3).

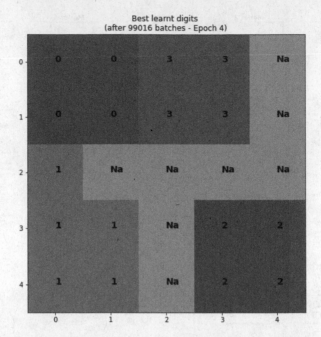

Fig. 4. Networks maps with the best-learned digits. For each classifier, we calculated the validation accuracy of every digit and we reported on a map the digit with highest accuracy. For each classifier, only digits with an accuracy superior to 0.8 are depicted on the maps. Digits with an accuracy below 0.8 are depicted as 'Na'.

4   Conclusions and Outlook

We presented a ML system in which the learning of a complex classification task is divided among different subsystems (neural network classifiers). The division of labor in learning the different subtasks is a self-organization process according to competitive and collaborative rules, inspired by self-organizing maps.

We showed that the system was able to correctly classify MNIST digit data, with each subsystem specializing for a particular digit. Through training, the neural network classifiers organize themselves and constitute spatial clusters of networks specialized in recognizing the same digits.

We observed that classifiers between clusters of specialized classifiers were not able to specialize for a single digit. The lack of performance may be explained by the Moore neighborhood function that pulls the classifiers between two clusters into two different learning directions. This process may paralyze such classifiers. Increasing the size of the networks map may therefore reduce the proportion of non-learning networks.

In the case of a pure top-down learning model, a failure of the main network causes the inability to accomplish the learning process. In contrast, a bottom-up approach based on the division of labor between different classifiers brings much more robustness. Indeed, in case of a failure of one subsystem, we could observe a compensation by the neighboring subsystems. This principle occurs frequently in the brain to compensate for neuronal loss after brain injury [9]. In the context of our general model, the non-learning networks could additionally constitute a reservoir of networks in case some other networks become defective. Further analysis would be needed to demonstrate that the knockdown of some networks leads to compensatory mechanisms by the non-learning networks.

In conclusion, this preliminary study serves as a proof of principle for our bottom-up system with neural network classifiers. To further demonstrate the usefulness and robustness of the system, we need to investigate possible extensions and use other types of neural networks such as autoencoders. Ultimately, these studies may be useful for the design of more complex adaptive autonomous systems [10].

References

1. Gerstner, W., Kistler, W. M., Naud, R., Paninski, L.: Synaptic plasticity and learning. neuronal dynamics: from single neurons to networks and models of cognition. Cambridge University Press. ISBN 978-1107635197 (2014)

2. Ladyman, J., Lambert, J., Wiesner, K.: What is a complex system? Eur. J. Philos. Sci. **3**, 33–67 (2013)

3. Wandell, B. A., Dumoulin, S. O., Brewer, A. A.: Visual field maps in human cortex. In: Neuron (2007)

4. Kohonen, T.: Self-organized formation of topologically correct featrure maps. Biol. Cybernet. **43**, 59–69 (1982)

5. Kalteh, A.M., Hjorth, P., Berndtsson, R.: Review of the self-organizing map (SOM) approach in water resources: analysis, modelling and application. Environ. Model. Softw. **3**(7), 835–845 (2008)

6. Liu, Y., Weisberg, R.H.: A Review of Self-Organizing Map Applications in Meteorology and Oceanography. Self Organizing Maps - Applications and Novel Algorithm Design, Josphat Igadwa Mwasiagi, IntechOpen (2011). https://doi.org/10.5772/13146

7. Fyfe, C.: Topographic maps for clustering and data visualization. In: Fulher, J., Jain, L.C. (Eds.), Computational Intelligence: A Compendium. Studies in Computational Intelligence, vol 115. Springer, Berlin, Heidelberg (2008). https://doi.org/10.1007/978-3-540-78293-3_3

8. Gygax, G., Füchslin R.M., Ott, T.: Self-organized division of labor in networks of forecasting models for time series with regime switches. In: 2020 International Symposium on Nonlinear Theory and Applications, Proceedings, pp. 278–281 (2020)

9. Hylin, M.J., Kerr, A.L., Holden, R.: Understanding the mechanisms of recovery and/or compensation following injury. In: Neural Plasticity (2017)

10. Pfeiffer, R., Lungarella, M., Fumiya, I.: Self-organization, embodiment, and biologically inspired robotics. Science **318**, 1088–1093 (2007)

Evolutionary Music: Statistical Learning and Novelty for Automatic Improvisation

Mattia Barbaresi[✉] and Andrea Roli

Dipartimento di Informatica - Scienza e Ingegneria, Alma Mater Studiorum
Università di Bologna, Campus of Cesena, Via dell'Università 50, Cesena, Italy
{mattia.barbaresi,andrea.roli}@unibo.it

Abstract. In this work we combine aspects of implicit learning with
novelty search in an evolutionary algorithm with the aim to automatically generate melodies with improvisational flavour. Using Markov
chains, the technique we present combines implicit statistical knowledge,
extracted from musical corpora, with an adaptive novelty search mechanism. The algorithm is described along with the main design choices.
Preliminary results are shown in two different musical contexts: Irish
music and counterpoint compositions.

Keywords: Evolutionary art · Computational creativity · Statistical
learning · Music · Novelty

1 Introduction

Computational Creativity (CC) is a renewed and vivid field of AI research that
aims at understanding human creativity while trying to produce "machine creativity", in which the computer appears to be creative, to some degree [7].
Leaning on such general definitions, approaches to CC often have a twofold
perspective: *(i)* developing systems that generate "creative" artefacts and *(ii)*
take this opportunity to investigate the cognitive aspects of such processes on
a computational basis. Following such practice, this work builds on the statistical approach to implicit human learning and cognitive development, aiming at
engineering more human-like and creative computational procedures.

Implicit Statistical Learning (ISL) refers to the general, implicit and ubiquitous ability of the brain to encode temporal and sequential phenomena and
more generally, to grasp the regularities in the environment, in an implicit and
unconscious way. This approach results from the recent attempt to unify two
research venues in psychology and cognitive science, namely Implicit Learning
(IL) and Statistical Learning (SL) [9, 26]. Implicit Learning refers more in general to mechanisms and knowledge, in the brain, that are unconscious. Statistical
Learning, on the other hand, was initially introduced for language acquisition,
and it is now invoked in various domains of psychology and neuroscience to
account for the human ability to detect and use statistical regularities present in

© The Author(s), under exclusive license to Springer Nature Switzerland AG 2022
J. J. Schneider et al. (Eds.): WIVACE 2021, CCIS 1722, pp. 172–183, 2022.
https://doi.org/10.1007/978-3-031-23929-8_17

the environment [30]. Additionally, many animal species are sensitive to distributional statistics, which suggests that learning from distributional statistics is a domain-general ability rather than a language-specific one [2]. More remarkably, it has been suggested that improvisational musical creativity is mainly formed by implicit knowledge. The brain models music—and other sequential phenomena such as language or movements—as a hierarchy of dynamical systems encoding probability distributions and complexity [14]. SL also plays a role in the production of sequences (e.g. notes or actions); from a psychological perspective, transitional probabilities distributions (TPs) sampled from music may refer to the characteristics of a composer's implicit (statistical) knowledge: a high-probability transition may be one that a composer is more likely to predict and choose, compared to a low-probability transition corresponding more to an unusual variation [12].

Based on these assumptions, this work aims at combining implicit-knowledge mechanisms with novelty search in a genetic algorithm to emulate an agent's (i.e. the musician who is composing impromptu) effort to produce novel and interesting sequences of actions (musical pieces) which have to be, at the same time, both familiar (concerning the knowledge initially provided) and novel. We assess our technique in two musical contexts that are characterized by a high degree of improvisation: Irish music and counterpoint.

2   Related Work

The applications of Markov chains in music have a long history dating back to the 1950s [27]: for detailed reviews on AI methods in music, or other examples and techniques, see [8,16,19,22,23]. Similarly, evolutionary computation has been used for generating music since long [5,20] and there are currently several systems that generate music by means of an evolutionary technique [4,22,24].

From our perspective, however, music generation is just a case study: we focus on modeling a general (context-independent) method for generating sequences (not limited to music) based on implicit mechanisms. In addition, the search towards creativity represents a different approach compared to the more common optimization practice, as the objective function tries to capture several, somehow subjective, features of the piece of art produced.

However, some of the latest and most comparable approaches to this work are perhaps those in [15,21,25]. *Continuator* is an interactive music performance system that accepts partial input from a human musician and continues in the same style as the input [25]. The system utilizes various Markov models to learn from the user input. It tries to continue from the most-informed Markov model (higher order), and if a match is not found with the user input, the system continues with the less-informed ones. In [15] the authors describe a method of generating harmonic progressions using case-based analysis of existing material that employs a variable-order Markov model. They propose a method for a human composer to specify high-level control structures, influencing the generative algorithm based on Markov transitions. In [21] the authors propose to capture phrasing structural information in musical pieces using a weighted variation of a first-order Markov chain model. They also devise an evolutionary procedure that tunes these weights for tackling the composer identification task between two composers. Another work is that of *GenJam* [5]. It uses a genetic algorithm to generate jazz improvisations, but it requires a human to judge the quality of evolved melodies. Finally, *GEDMAS* is a generative music system that composes entire Electronic Dance Music (EDM) compositions. It uses first-order Markov chains to generate chord progressions, melodies and rhythms [1]. In a previous work [3] we conceived a genetic algorithm (GA) which, starting from a given inspiring repertoire and a set of unitary moves, generates symbolic sequences of movements (i.e. choreographies) exploiting similarity with the repertoire combined with the novelty search approach [35].

3   Materials and Methods

According to blind-variation and selective-retention principles of creativity, creative ideas must be generated without full prior knowledge of their utility values [32, 33]. Herein, genetic algorithms offer a natural setting for the blinded-divergent and convergent mechanisms involved in the creative process, terms of diversification and intensification [6]. The evolutionary approach is in itself an exploratory process: the combination of two individuals from the population pool is a combinational process, but the use of a fitness function guides the exploration toward promising areas of the conceptual space, which is bounded and defined by the genetic encoding of the individuals. Losing the fitness function, or having one that is unable to effectively guide the exploration, reverts the mechanism to pure combinational creativity, where elements of the conceptual space are joined and mutated hoping to find interesting unexplored combinations. In this work, we combine an adaptive genetic algorithm with Markov chains—built up from a corpus of music excerpts. The algorithm evolves the parameters (weights) of a constructive procedure that acts on the model and produces new pieces of music that are intended to be novel variations upon familiar music. In addition, the model is used also for evaluating the similarity (the objective function) of generated sequence to the starting knowledge. In this work we build upon the novelty approach presented in [3] and we make it adaptive so as to make it independent on specific ranges of the functions involved in the algorithm. Algorithm 1 shows the main loop and Algorithm 2 shows such a procedure.

3.1   Markov Model: Chains and Score

Markov chains allow us to grasp the statistical structure of sequential phenomena (i.e. music, movements) but also statistical learning and knowledge in humans [13]. It has been observed that transitional probabilities sampled from music (based on Markov models) may also refer to the characteristics of a musician's statistical knowledge and captures temporal individual preferences in

---

Algorithm 1: Pseudo code for GA

$eval \leftarrow$ mono
for *#iters* **do**
    $eval$(pop)
    offspring, elite $\leftarrow$ pop
    offspring crossover and mutation
    pop $\leftarrow$ offspring + elite
    $eval \leftarrow select\_objective$(pop)
    $archive \leftarrow archive\_assessment$(elite)

Algorithm 2: Pseudo code for *select_objective*() function

$$eval = \begin{cases} mono & \text{// the Markov Score} \\ biobjective & \text{// Pareto(MarkovScore(), Novelty())} \end{cases}$$

$prevFit = bestFit$
$bestFit = selBest$(pop)
if $eval == mono$ **then**
    if $prevFit \simeq bestFit$ **then**
        counter $\leftarrow$ counter - 1
        if $counter == 0$ **then**
            $reset$(counter)
            $lastAvg = avg$(pop)
            $eval \leftarrow biobjective$
    else
        $restart$(counter)
else
    $bestAvg = avg$(pop)
    if $lastAvg \simeq bestAvg$ **then**
        $eval \leftarrow mono$

---

improvisation [11]. We consider here sequences of symbols from a finite alphabet, which can represent e.g. melodies. To model this implicit knowledge, we computed the Markov chains with memory m (or Markov chain of order m) up to the $m = 5$ order[1] starting from a set of musical pieces. For each order m, transitional probabilities are computed for each excerpt as frequency ratios: $P(y|x_m) = \frac{\#x_m y}{\#x_m}$, given a symbol y and a (sub-)sequence x_m (the past) of length m, where $x_m \rightarrow y$ is the inspected transition. As we increase the context size, the probability of the alphabet becomes more and more skewed, which results in lower entropy. The longer the context, the better its predictive value.

---

[1] In the data we used, orders higher than 5 are not "expressive" because of the data limits and structure: at some point, higher orders contain about the same information held in the previous ones.

3.2   Objective Function

The objective function is intended to capture the familiarity (or the membership, the similarity) of a sequence with respect to the Markov model resulted from the (inspirational) musical corpus. So given a sequence $X = x_0 x_1 ... x_n$ the Markov score is defined as the product of TPs of symbols in the sequence

$$score(X) = P(x_0) \times P(x_1|x_0) \times P(x_2|x_0 x_1) \times ... \times P(x_n|x_{n-m}...x_{n-1}) \quad (1)$$

where $x_{n-m}...x_{n-1}$ is the (past) sequence of length m up to the $(n-1)_{th}$ symbol. For a given chain, it might happen that a transition (past $\rightarrow$ symbol) does not exist. If that actual past does not match a transition in that current order, we shorten the past ($x_{n-m}...x_{n-1}$ becomes $x_{n-m+1}...x_{n-1}$) and move down a order (i.e. a chain), looking at shorter contextual information to guide the generation. Finally, we apply the negative logarithm to the Markov score and turn the GA objective into a minimization problem:

$$\underset{X}{minimize}: \quad -\log(score(X)) \quad (2)$$

3.3   Encoding

The GA manipulate the parameters of a randomized constructive procedure that acts on the Markov model. The genotype is an array of 6 decimal elements— that sum up to 1—representing the weights to assign to each computed chain in the Markov model. Namely a weight i for each i_{th}-order Markov chain (i.e., a categorical distribution for the chain choice), as in Table 1. Every positional value of the array weights the probability of the corresponding order in the model when generating a sequence. Notably, these arrays weigh a Monte Carlo process that selects, for each symbol to be emitted in the generation, the order–i.e. the Markov chain–to look at when looking for the transitions to produce it. The phenotype indeed is represented by all sequences generated by the model with that given array of weights (the individual of the GA).

Table 1. Example of an individual

w_0	w_1	w_2	w_3	w_4	w_5
0.0	0.3	0.05	0.4	0.2	0.05

3.4   Adaptive Novelty Search

To steer the generation towards novel productions, we followed the novelty search described in [3,35]. This algorithm is applied to compensate for a lack of diversity concerning the best individuals already found. It consists of a bi-objective optimization activated when the main objective (e.g. the fitness) stagnates. The novelty is computed as the mean L_2 norm between a genotype (weight vector) and an archive of past genotypes. Using the Markov model, there is no explicit boundary for the objective function since the Markov score depends on the length of the sequence being evaluated. To tackle this problem, we conceived an adaptive mechanism for the activation of novelty based on the results obtained in previous iterations. At each iteration, the algorithm stores the best result obtained. If such value does not change for a given number of iterations, somehow the algorithm is stuck at a minimum. In such a case, the algorithm starts to look at the novelty of individuals. When novelty search has moved the score away of a certain amount from the last best value found, it is turned off and the regular evolution with the Markov score is resumed. See Eq. 3.

$$\begin{cases} \text{if } bestFit \simeq prevFit, \text{for } k \text{ times,} & \text{switch to bi-objective} \\ \text{if } |lastAvg - bestAvg| \simeq stdevLast, & \text{switch to mono} \end{cases} \quad (3)$$

As well as for the objective function, we applied the negative logarithm to novelty too. Thus the bi-objective optimization is intended to minimize both the main objective and the novelty of individuals.

$$\underset{X}{minimize}: \quad -\log(novelty(X)) \quad (4)$$

We remark that biasing towards novelty does not mean just adding randomness, but rather diversifying with respect to the best solutions found.

Archive Assessment. For the assessment of the archive we followed the approach used in [35] except for one aspect; we did not consider a threshold for the individual in order to be added to the archive. Instead we considered, at each iteration of the genetic algorithm, the individuals of the elite, from the elitism process. For these individuals we calculate the dissimilarity as in the mentioned work.

4   Results

The most suitable musical contexts in which our technique can be applied are those in which improvisation plays an important role; but we also need structure to some degree, in such a way that the implicit (soft) constraints imposed by the style can be detected. This way, the music resulting from our method has some amount of novelty, yet still in the style of the examples provided. For our experiments we chose two notable musical contexts: traditional Irish tunes and Orlande de Lassus' *Bicinia* [18]. In this section we first introduce these cases and subsequently we present a selection of the typical results achieved by our technique.

4.1   Irish Songs and Bicinia

Traditional Irish music is strongly characterized by its melodies: most old tunes are just melodic (see e.g. [34]) or they are the result of an improvisation upon a given *ground*, i.e. a bass line providing also a harmonic base (see e.g. [28]). In any case, the melodic part of a traditional Irish music is currently the most important component and melodies are usually played with variations, improvising upon a given melodic structure. A large corpus of traditional Irish airs is available in abc notation,[2] which makes it possible to extract melodies as sequences of symbols, each representing both note and duration. A typical traditional Irish air is shown in Fig. 1. From these airs we extracted all the ones in the key of G and assigned one symbol to each ⟨*note,duration*⟩ pair.

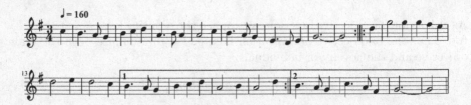

Fig. 1. Score of a well known traditional air titled "The south wind".

The second musical context we have chosen is that of two voices counterpoint, which is one of the simplest and oldest forms of polyphony [29]. In origin, a voice was superimposed to a given one, called *cantus firmus*, in improvisational settings. This original impromptu spirit was subsequently substituted by a more elaborated compositional approach, leading to marvelous multi-voices counterpoints, such as the ones composed by Gesualdo da Venosa. The main technical characteristic of these pieces of music can be summarized in a small set of rules involving the intervals, i.e. the distance in semitones, between the upper and the lower voice. For example, the distance between C and F (above C) is 5 semitones. Obviously, these are not all hard constraints, but some are rather preferences, and they have also been subject to change in time according to different musical aesthetics. Examples of such rules, typical of XVI century counterpoint, are:

- no parallel fifths or octaves are allowed (e.g. C-G cannot move to D-A)
- fifths and octaves should be intercalated by *imperfect consonances*, i.e. thirds and sixths (e.g. an allowed sequence is C-G, C-A, D-A)
- *dissonances*, i.e. all intervals except for unisons, octaves and fifths, should *be prepared* and then *resolve* to a consonant interval by descending (e.g. D-B, C-B, C-A)

---

[2] http://www.norbeck.nu/abc/.

Fig. 2. Excerpt of bicinium no. 1, "Te deprecamus", by de Lassus. Score extracted from https://imslp.org/wiki/Category:Lassus,_Orlande_de.

In our tests, we have taken all the twelve two-voices counterpoint compositions, called *bicinia*, by Orlande de Lassus, which are available as MIDI files.[3] In Fig. 2 we show an excerpt of a bicinium by de Lassus. This second context was chosen to assess to what extent our method is able of identifying recurrent patterns and rules typical of a music genre. In this case, we have encoded the twelve MIDI bicinia as sequences of intervals (i.e. distances in semitones between upper and lower voice). As the two voices have in general different durations, we have sampled the music at steps of duration 1/32 and taken the intervals in semitones, deleting repetitions. This provides the repertoire on which the Markov models are computed. A typical result from our system is a sequence of integer numbers representing intervals in semitones which can be used as a guideline for composing the upper voice upon a given *cantus firmus*.

4.2   Experimental Settings

Differently from usual optimization contexts, in our case a good performance does not correspond to the one that leads to the overall best objective function values, but rather to a good balance between similarity (Markov score) and novelty. Therefore, we tuned the parameters of the algorithm trying to attain an effective interplay between score and novelty. The results we present have been obtained with a population of 100 individuals, uniform crossover with probability 0.5, Gaussian mutation ($\mu = 0$, $\sigma = 0.3$) with both chromosome and gene probability equal to 0.35, and 200 generations. The novelty is activated after 5 idle generations (the best score s_{pop} in the current population is stored, along with the standard deviation of the populations scores σ_{pop}) and deactivated when the difference between the score of current best individual and s_{pop} is greater than $\sigma_{pop}/3$. The plot of score and novelty of a typical run is shown in Fig. 3.

We can observe that the score oscillates: whenever the algorithm stagnates, novelty is activated so as to increase diversification. When this latter is high enough, only the Markov score is kept as objective function. In a sense, we can describe the dynamics of the algorithms as a biased exploration of local minima, as typically done by Iterated Local Search techniques [6].

---

[3] http://icking-music-archive.org/ByComposer1/Lasso.php.

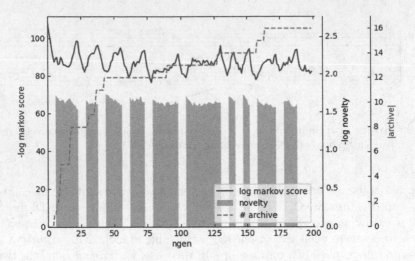

Fig. 3. Plot of score and novelty of a typical run. Both the functions are to be minimized and novelty is activated, adaptively, only when diversification is needed. The number of individuals in the archive, involved in the calculation of novelty [3], is also plotted.

4.3   Musical Results

Due to limited space we can just provide a few examples of the musical results obtained. The generation of melodies inspired to traditional Irish airs has been evaluated by sampling some weight vectors from the final populations and using them to generate actual music. By analyzing the results both through visual inspection and by listening to them, we observed that the music generated is similar to the repertoire provided but with variations and recombinations of patterns. A couple of excerpts are shown in Fig. 4, where we can observe variations of typical Irish melodic and rhythmic patterns: the characteristic *run* (i.e. a fast sequence of notes, typically in a scale) in bars 4, 5 of the first example and the syncopated and composite rhythm in the second one.

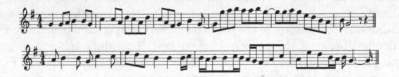

Fig. 4. Two typical excerpts of automatically created Irish music.

The second test case concerns de Lassus' Bicinia. The main result attained is that it was able to discover the basic rules that characterize two voices counterpoint. In particular the rules extracted that have more strength are: *incipit* with a perfect consonance (unison, octave or fifth), no consecutive octaves or fifths, and dissonant intervals followed by consonant ones—both perfect and imperfect. In Fig. 5 we show an excerpt of the counterpoint produced by applying one of the sequences generated by our algorithm to a given cantus firmus (Chanson CXXVI from manuscript Bibl. Nat. Fr 12744 published by G. Paris). As the algorithm returns a sequence of intervals, it could be used as a tool that assists composers by suggesting feasible note choices, once one of the two voices (*cantus firmus*) is given.

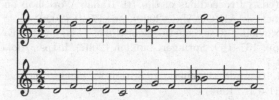

Fig. 5. An example of a two voices counterpoint. The lower voice is the cantus firmus, while the upper voice has been generated by applying a sequence of intervals generated by our algorithm.

In conclusion, for both the contexts the algorithm was able to identify the core regularities and to elaborate around them. The calibration of the parameters is important to achieve a good balance between the tendency of just recombining the patterns learned and the exploration of new possibilities. However, the choice of parameter values does not seem critical, because the combined use of a stack of Markov models of varying orders and novelty search makes it easier to achieve this trade-off.

5   Conclusion and Future Work

The algorithm we have presented has proven to be able to generate novel, yet somehow familiar melodies. An interesting perspective is that of creating non-homogeneous repertoires, maybe just including music that the user likes, without any genre restriction. This way, our system can produce music that merges some of the peculiar features that meet user's tastes.

Future work is focused on quantitatively assessing the properties of the generated sequences by means of information theory measures, such as block entropies [31] and complexity measures like set-based complexity [17]. Some of these measures can also be introduced in the generative process, so as to limit human evaluation as much as possible. In addition, some metrics can also be used to assess the distance between sequences or to cluster them [10].

As the proposed technique is general and can be applied whenever the goal is to produce sequences of actions, we plan to explore multimodal automatic generation by combining Markov models from two different contexts, e.g. music and text.

References

1. Anderson, C., Eigenfeldt, A., Pasquier, P.: The generative electronic dance music algorithmic system (gedmas). In: Proceedings of the AAAI Conference on Artificial Intelligence and Interactive Digital Entertainment (2013)
2. Armstrong, B.C., Frost, R., Christiansen, M.H.: The long road of statistical learning research: Past, present and future (2017)
3. Barbaresi, M., Bernagozzi, S., Roli, A.: Robot choreographies: Artificial evolution between novelty and similarity. In: Finzi, A., Castellini, A., Buoncompagni, L., Anzalone, S. (eds.) Proceedings of the 7th Italian Workshop on Artificial Intelligence and Robotics (AIRO@AIxIA2020), pp. 17–21 (2020)
4. Biles, J.: Improvizing with genetic algorithms: Genjam. In: Evolutionary Computer Music, pp. 137–169. Springer, London (2007). https://doi.org/10.1007/978-1-84628-600-1_7
5. Biles, J., et al.: Genjam: A genetic algorithm for generating jazz solos. In: ICMC, vol. 94, pp. 131–137 (1994)
6. Blum, C., Roli, A.: Metaheuristics in combinatorial optimization: Overview and conceptual comparison. ACM Comput. Surv. 35(3), 268–308 (2003)
7. Boden, M.A.: Creativity in a nutshell. Think 5(15), 83–96 (2007)
8. Carnovalini, F., Rodà, A.: Computational creativity and music generation systems: An introduction to the state of the art. Front. Artif. Intell. 3, 14 (2020)
9. Christiansen, M.H.: Implicit statistical learning: A tale of two literatures. Top. Cognit. Sci. 11(3), 468–481 (2019)
10. Cilibrasi, R., Vitányi, P.: Clustering by compression. IEEE Trans. Inf. theory 51(4), 1523–1545 (2005)
11. Daikoku, T.: Musical creativity and depth of implicit knowledge: Spectral and temporal individualities in improvisation. Front. Comput. Neurosci. 12, 89 (2018)
12. Daikoku, T.: Neurophysiological markers of statistical learning in music and language: Hierarchy, entropy and uncertainty. Brain Sci. 8(6), 114 (2018)
13. Daikoku, T.: Time-course variation of statistics embedded in music: Corpus study on implicit learning and knowledge. PLoS One 13(5), e0196493 (2018)
14. Daikoku, T.: Depth and the uncertainty of statistical knowledge on musical creativity fluctuate over a composer's lifetime. Front. Comput. Neurosci. 13, 27 (2019)
15. Eigenfeldt, A., Pasquier, P.: Realtime generation of harmonic progressions using controlled Markov selection. In: Proceedings of ICCC-X-Computational Creativity Conference, pp. 16–25 (2010)
16. Fernández, J.D., Vico, F.: Ai methods in algorithmic composition: A comprehensive survey. J. Artif. Intell. Res. 48, 513–582 (2013)
17. Galas, D., Nykter, M., Carter, G., Price, N., Shmulevich, I.: Biological information as set-based complexity. IEEE Trans. Inf. Theory 56(2), 667–677 (2010)
18. Haar, J.: Lassus, orlande de (2021). https://doi.org/10/g5nj, Grove Music Online, accessed on 12 Nov 2021
19. Herremans, D., Chuan, C.H., Chew, E.: A functional taxonomy of music generation systems. ACM Comput. Surv. 50(5), 1–30 (2017)
20. Horner, A., Goldberg, D.: Genetic algorithms and computer-assisted music composition, vol. 51. Michigan Publishing, University of Michigan Library, Ann Arbor (1991)
21. Kaliakatsos-Papakostas, M.A., Epitropakis, M.G., Vrahatis, M.N.: Weighted Markov chain model for musical composer identification. In: Di Chio, C., et al. (eds.) EvoApplications 2011. LNCS, vol. 6625, pp. 334–343. Springer, Heidelberg (2011). https://doi.org/10.1007/978-3-642-20520-0_34

22. Liu, C.H., Ting, C.K.: Computational intelligence in music composition: A survey. IEEE Trans. Emerg. Top. Comput. Intell. **1**(1), 2–15 (2016)
23. Mor, B., Garhwal, S., Kumar, A.: A systematic literature review on computational musicology. Archiv. Comput. Methods Eng. **27**(3), 923–937 (2020)
24. Muñoz, E., Cadenas, J., Ong, Y., Acampora, G.: Memetic music composition. IEEE Trans. Evolution. Comput. **20**(1), 1–15 (2014)
25. Pachet, F.: Interacting with a musical learning system: The continuator. In: Anagnostopoulou, C., Ferrand, M., Smaill, A. (eds.) ICMAI 2002. LNCS (LNAI), vol. 2445, pp. 119–132. Springer, Heidelberg (2002). https://doi.org/10.1007/3-540-45722-4_12
26. Perruchet, P., Pacton, S.: Implicit learning and statistical learning: One phenomenon, two approaches. Trends Cognit. Sci. **10**(5), 233–238 (2006)
27. Pinkerton, R.C.: Information theory and melody. Sci. Am. **194**(2), 77–87 (1956)
28. Playford, J.: The Division Violin: Containing a Collection of Divisions Upon Several Grounds for the Treble-Violin, 3rd edn. Henry Playford, London (1688)
29. Sachs, K.J., Dahlhaus, C.: Counterpoint (2001). https://doi.org/10/g5nk, Grove Music Online. Accessed 12 Nov 2021
30. Saffran, J.R., Kirkham, N.Z.: Infant statistical learning. Annu. Rev. Psychol. **69**, 181–203 (2018)
31. Shannon, C.: A mathematical theory of communication. Bell Syst. Tech. J.**27**(1–2), 379–423, 623–656 (1948)
32. Simonton, D.K.: Creativity as blind variation and selective retention: Is the creative process darwinian? In: Psychological Inquiry, pp. 309–328 (1999)
33. Simonton, D.K.: Defining creativity: Don't we also need to define what is not creative? J. Creativ. Behav. **52**(1), 80–90 (2018)
34. Neal, J., Neal, W.: A Collection of the Most Celebrated Irish Tunes Proper for the Violin, German Flute or Hautboy, Dublin, Ireland (1724)
35. Vinhas, A., Assunção, F., Correia, J., Ekárt, A., Machado, P.: Fitness and novelty in evolutionary art. In: Johnson, C., Ciesielski, V., Correia, J., Machado, P. (eds.) EvoMUSART 2016. LNCS, vol. 9596, pp. 225–240. Springer, Cham (2016). https://doi.org/10.1007/978-3-319-31008-4_16

Semantic Search

ARISE: Artificial Intelligence Semantic Search Engine

Luigi Di Biasi[1,2] , Jacopo Santoro[2], and Stefano Piotto[2,3(✉)]

[1] Department of Computer Science, University of Salerno, Via Giovanni Paolo II,
132, 84084 Fisciano, Italy
[2] Department of Pharmacy, University of Salerno, Via Giovanni Paolo II, 132,
84084 Fisciano, Italy
piotto@unisa.it
[3] Bionam Center for Biomaterials, University of Salerno, Via Giovanni Paolo II, 132,
84084 Fisciano, Italy

Abstract. Thanks to the services provided by the major cloud computing providers, the rise of Artificial Intelligence (AI) appears to be inevitable. Information analysis and processing, where the primary purpose is to extract knowledge and recombine it to create new knowledge, is an interesting research topic where AI is commonly applied. This research focuses on the semantic search problem: semantic search refers to the ability of search engines to evaluate the intent and context of search phrases while offering content to users. This study aims to see if introducing two biologically inspired characteristics, "weighting" and "correlated" characters, may increase semantic analysis performance. First, we built a preliminary prototype, ARISE, a semantic search engine using a new Artificial Network architecture built upon a new type of Artificial Neuron. Then, we trained and tested ARISE on the PubMed datasets.

Keywords: Semantic search · Artificial Intelligence · Neural Networks

1 Introduction

The rise of AI seems to be unstoppable thanks to the services offered by the majors' cloud computing providers. Researchers and private companies can study and deploy their own AI systems to solve their problems without designing and implementing their physical computing architecture. In addition, many packages were released to allow the researchers to focus on the issue to solve without fully knowing the underlying tools and techniques.

Information analysis, sentiment analysis, and information extraction and processing are three interesting topics where AI is commonly used in the design of AI-based applications. The primary goal in these fields is to extract information from a dataset (knowledge base) and recombine it to produce new knowledge.

This work is focused on the semantic search issue. When providing content to users, semantic search refers to the ability of search engines to evaluate the intent and context of search words. In lexical search, the search engine seeks literal

J. J. Schneider et al. (Eds.): WIVACE 2021, CCIS 1722, pp. 187–192, 2022.
https://doi.org/10.1007/978-3-031-23929-8_18

matches of the query words or variants without understanding the question's overall meaning.

In this research, we aim to investigate if two biological-inspired aspects, the "weighting" and "correlated" characters by Felsenstein [3] can be used to design a new kind of Neural Network architecture focused on semantic analysis problem. We temporary identify this architecture with the name MEF-Based Associative Neural Network - MEFNET. Also, we are proposing an alternative learning methodology not related to the *train-validation-test approach* but on the more biological *penalty or reward* concept.

The results in [1] suggests that "the evolutionary process takes place at different scales simultaneously". Also, the results in [5] suggest that the "common sub-portion of a biological entity" can maintain its functionality even after long periods of evolution, even after replacements and permutations have occurred. Thus, a common sub-portion (common patterns) between two entities could determine common observable functions, regardless of the common pattern position. Following this evidence, we assumed as a working hypothesis that *"it is not the common pattern determining entities' similarity but the number of pattern occurrences."*; in our case, this means that two entities could be very similar even if the important common parts are placed in different areas.

Due to the previous observation and hypothesis, we defined another working idea: *"if two entities show many functional parts in common, then these entities are functionally very similar."* Therefore, we called the most expressed functional part of entities the *Most Expressed Features* (MEF).

In our experimentation, we used the entire PubMed dataset as the MEFNET learning dataset. We considered each Pubmed Papers as a biological entity. The content of each paper (title and abstract) is regarded as the proteome of the object expressed in strings form.

The next section reports the preliminary results we obtained with MEFNET. Finally, we build an initial prototype for a semantic search engine called Arise[1] using the MEFNET and the entire PubMed datasets.

2   Related Works

In 2009, Sims et al. [9] used a profiles frequency approach to compare different species genomes. In 2017, Ferraro et al. [4] exploited alignment-free - AF- methodology to compare FASTQ sequences. Also, still in 2017, Cattaneo et al. [2] implemented a map/reduce system with Hadoop to allow AF analysis of big biological sequences dataset. All these works tried to defined metrics to compare biological entities. In 2019, Nardiello et al. [6] described a pseudo-semantic analysis approach applied to molecular dynamics. In 2019, Piotto et al. [8] suggested a possible universal coding of materials by reporting the suggestion that pseudo-semantic analysis may be particularly advantageous in the study of polymeric biomaterials.

---

[1] https://www.softmining.it.

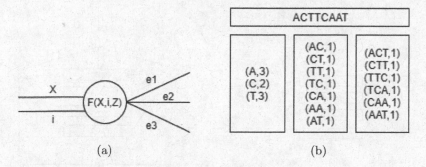

Fig. 1. The MEFNET base neuron (a) and a MEF table computed on a simple string (b).

3 Materials and Methods

Following our working hypothesis, we assumed that entities could be compared using their MEF. Therefore, we use the MEF extracted from each Pubmed paper to build our MEFNET.

3.1 Defining MEF

The MEF can be thought of as object tags, as Fig. 1(b) shows. It is strictly related to n-gram (and k-mer) concepts: an n-gram is a generic string with length n; a k-mer is a generic string of length k. Thus, it is possible to use n-gram or k-mer as synonymous.

As an example, let be the ACTTCAAT string an entity. Then, using a classic multi-sliding-window word-count approach, it is possible to perform the feature extraction step by extracting both n-grams and occurrences. Our proposed system needs to use the *most expressed features* due to the working hypothesis. Then, the occurrences were sorted in descendant order at the end of the extraction. In our experimentation, we extracted the MEF for each Pubmed Paper.

3.2 Artificial Neuron Structure

In MEFNET, we use a slight perceptron modification, as shown in Fig. 1(a).

Each neuron has two inputs (X, i) and multiple outputs, each of them identified by the edge value $(e_1, \ldots, , e_n)$. X and Z represents a n-gram while i is an integer value. The activation function $F(X, i, Z)$ will fire the edge e_i if and only if $i <= e_i$ and $X = Z$.

3.3 MEFNET Learning Process

The neurons creation happens during the learning process that occurs during the MEFs extraction. Thus, the network does not have a fixed structure but can evolve following knowledge acquisition.

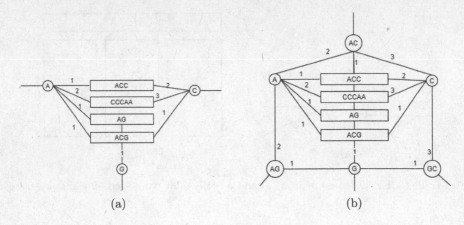

Fig. 2. A mef-nn with one layer (a) and with two layer (b) build from the same dataset.

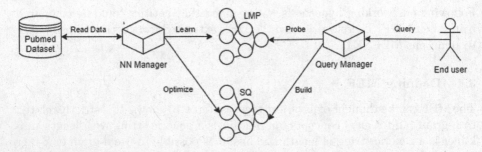

Fig. 3. The ARISE prototype architecture.

Figure 2 shows an example of the learning process of four simple objects. For simplicity, The alphabet of the object contains only three characters (ACG). In this example, we restricted the n-gram size to one. Then, MEFNET contains only three neurons: each delegate to learn information for a single n-gram for multiple learned entities. If we execute the same example on the Pubmed dataset, MEFNET will contain only 22 neurons with millions of outputs because the outputs are not upper-bounded. For optimization reasons, MEFNET allows multiple layers. The figure shows the same network with two-layer. The number of layers is related to the maximum n-gram permitted length.

In our preliminary experiments, we followed an unsupervised learning approach. Therefore, we could not define a priori the standard range (min and max sliding window size) to drive MEF extraction and network design. Instead, we leave the network free to evolve driven by the PubMed dataset. To extract MEF, store it, build and execute the network, we used a standard map/reduce approach by deploying a GRIMD [7] instance on Amazon AWS.

3.4 Query Execution

At the current development stage, ARISE is composed of two parallel MEFNET.

The first one, LMP, maintains the original results of the learning step. Hence, this network can retrieve potential candidates regarding user queries. Therefore, LMP can be considered a smart in-memory multilevel index. The second one, SQ, grew following ARISE utilization during the time.

A query to ARISE will cause the activation of LMP and the creation of a new entry in SQ. In particular, defined Q as our query, the execution of the action $Probe(Q)$ will extract the MEF from Q. The extracted MEF is used as an activator of the LMP to retrieve potential output candidates. As a result, a set of possible outputs POT composed of all the entities linked to an edge activated is created. The sum of the edge values is used as a primary scoring function. Finally, the activated neuron from LMP (a subnetwork) expands the SQ network by a merge action.

The POT and the MEF extracted by $Probe(Q)$ are used to expand the SQ training sets. In addition, ARISE users can assign a penalty or a reward to each POT regarding the quality of the output. These values are used during SQ hyperparameters optimization. Figure 3 reports the entire prototype architecture.

4 Result and Conclusion

The proposed approach allowed ARISE to perform the query on the entire PubMed dataset, avoiding utilizing standard query operators (like, and, or, not, between). Thus, it can be considered a sort of AI DBMS. Also, the presence of the SQ and LMP network allows researchers to change MEF comparison metrics without needing to retrain the entire network. ARISE can provide the users a list of papers (identified by doi, title and abstract) correlated to the user query. In addition, each result is provided with a score computed by SQ, related to the MEF comparison metric used. In future work, a validation step for each metric is needed to understand better performance. Due to the high computational power needed (both in Running Time and in Space), multiple distributed approaches must be evaluated in the future.

References

1. Attwood, T.K.: The babel of bioinformatics. Science **290**(5491), 471–473 (2000)
2. Cattaneo, G., Petrillo, U.F., Giancarlo, R., Roscigno, G.: An effective extension of the applicability of alignment-free biological sequence comparison algorithms with hadoop. J. Supercomput. **73**(4), 1467–1483 (2017)
3. Felsenstein, J.: Numerical methods for inferring evolutionary trees. Quart. Rev. Biol. **57**(4), 379–404 (1982)
4. Ferraro Petrillo, U., Roscigno, G., Cattaneo, G., Giancarlo, R.: Fastdoop: A versatile and efficient library for the input of fasta and fastq files for mapreduce hadoop bioinformatics applications. Bioinformatics **33**(10), 1575–1577 (2017)

5. Koohy, H., Dyer, N.P., Reid, J.E., Koentges, G., Ott, S.: An alignment-free model for comparison of regulatory sequences. Bioinformatics **26**(19), 2391–2397 (2010)
6. Nardiello, A.M., Piotto, S., Di Biasi, L., Sessa, L.: Pseudo-semantic approach to study model membranes. In: Piotto, S., Concilio, S., Sessa, L., Rossi, F. (eds.) BIONAM 2019 2019. LNB, pp. 120–127. Springer, Cham (2020). https://doi.org/10.1007/978-3-030-47705-9_11
7. Piotto, S., Di Biasi, L., Concilio, S., Castiglione, A., Cattaneo, G.: Grimd: Distributed computing for chemists and biologists. Bioinformation **10**(1), 43 (2014)
8. Piotto, S., Nardiello, A.M., Di Biasi, L., Sessa, L.: Encoding materials dynamics for machine learning applications. In: Piotto, S., Concilio, S., Sessa, L., Rossi, F. (eds.) BIONAM 2019 2019. LNB, pp. 128–136. Springer, Cham (2020). https://doi.org/10.1007/978-3-030-47705-9_12
9. Sims, G.E., Jun, S.R., Wu, G.A., Kim, S.H.: Alignment-free genome comparison with feature frequency profiles (ffp) and optimal resolutions. Proc. Natl. Acad. Sci. **106**(8), 2677–2682 (2009)

Artificial Medicine and Pharmacy

Influence of the Antigen Pattern Vector on the Dynamics in a Perceptron-Based Artificial Immune - Tumour- Ecosystem During and After Radiation Therapy

Stephan Scheidegger[1]($\boxtimes$), Sergio Mingo Barba[1,2], Harold M. Fellermann[3], and Udo Gaipl[4]

[1] School of Engineering, Zurich University of Applied Sciences, Winterthur, Switzerland
scst@zhaw.ch
[2] University of Fribourg, Fribourg, Switzerland
[3] Interdisciplinary Computing and Complex Biosystems Research Group, School of Computing, Newcastle University, Newcastle Upon Tyne, UK
[4] Translational Radiobiology, Department of Radiation Oncology, Universitätsklinikum, Erlangen, Germany

Abstract. Artificial immune-tumor ecosystems can serve as models to explore the complex tumor-host-immune – interactions in silico. This may contribute to a better understanding of the conditions leading to anti-cancer immune response in patients during anti-cancer therapy. For model development, it is important to identify an appropriate model structure which is suitable to mimic the behavior of real biological systems. In this study, the influence of the number of antigens in an artificial adaptive immune system onto an immune-tumor ecosystem during and after radiation therapy (RT) is investigated. For antigen pattern recognition, a perceptron is used. The simulated scenarios with 4, 9 and 12 antigens exhibit differences in the immune response, but in all cases, perceptron weights for host tissue evolve after RT into negative values, leading to an immune-suppressive effect. This effect results from the evolution of the populations in the ecosystem and the training of the perceptron. In conclusion, the response of the proposed artificial immune system is strongly dependent on the ecosystem dynamics, which seems to be the case for the real biological systems as well.

Keywords: Systems medicine · Immune system in silico · Perceptron · Antigen pattern · Anti-cancer therapy · Adaptive immune response

1 Introduction

Theoretical, preclinical and clinical research demonstrated that radiation therapy (RT) is able to induce or to affect anti-tumor immune responses [1–3]. The idea of activating the immune system by local irradiation of a tumor leads to the question, whether and how RT could be used as a kind of an anti-tumor vaccine. The search of an answer may be supported by a profound understanding of the dynamic interplay between tumor, host tissue and the immune system. A tumor-host-immune system may be considered as an ecosystem [4–6].

The investigation of the dynamics in such systems would require an adequate model for the adaptive immune system. Artificial immune - tumor- ecosystems may be far away from a real patient, but they could serve as a laboratory to investigate fundamental principles of dynamics in such systems under well-controlled conditions [7]. As a complementary approach to biological experiments in vitro, in vivo or clinical trials, such "sandbox" games could be used to generate hypotheses supporting the design of clinical trials. Scheidegger et al. [8] proposed an artificial immune – tumor - model system covering two essential aspects: Ecosystem dynamics between host tissue and different tumor sub-clones and antigen pattern recognition by a learning (adaptive) immune system. The proposed model exhibited some interesting features: As a response onto radiation treatments, host tissue becomes immune-suppressive whereas the detection of tumor tissue by the immune system is improved under certain conditions.

In contrast to other mathematical models for immune-tumor systems [1, 9], we consider the adaptive immune system as a trainable (programmable) unit and anti-tumor treatments as means to train the immune system to battle against cancer. The purpose of this study is to investigate the effect of the number of the antigen pattern vector components onto the system response during and after RT, which has not been studied yet (in contrast to the influence of different branching in the mutation trees and to the sensitivity of the model onto different parameters [8]).

2 Materials and Methods

In this study, a modified version of the model proposed by Scheidegger et al. [8] is used. The model consists of two major components – a tumor ecosystem including host tissue and immune cells in the tumor compartment; and a perceptron [10] for antigen pattern recognition (see Fig. 1).

The idea of using a perceptron to mimic the immune system's ability of pattern recognition is based on the danger model proposed by P. Matzinger [11]. Following this concept, the immune system is only activated when a danger signal and antigens are coincidently present.

In the following, the model equations are presented (a detailed explanation of the model is given by Scheidegger et al. [8]). The dynamic interaction between the different tumor sub-clones T_{ik} and the host tissue H is given by the following system of ordinary differential equations:

$$\frac{dT_{11}}{dt} = (k_{T11} - k_{mut} - k_{eT} - r_{11}k_{IT} - k_{HT}H - k_{TT}T - (\alpha_T + 2\beta_T\Gamma) \cdot R) \cdot T_{11}$$

$$\frac{dT_{ik}}{dt} = (k_{Tik} - k_{eT} - r_{ik}k_{IT} - k_{HT}H - k_{TT}T - (\alpha_T + 2\beta_T\Gamma) \cdot R) \cdot T_{ik} + k_{mut} \cdot q_{il}T_{lk}$$

$$\frac{dH}{dt} = (k_{aH} - k_{eH} - r_H k_{IH} - k_{bH}H - k_{TH}T - (\alpha_H + 2\beta_H\Gamma) \cdot R) \cdot H$$

$$(1)$$

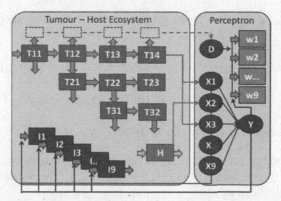

Fig. 1. Structure of the tumor – immune-system model: Bold arrows symbolize flows; thin arrows show the dependency of variables. Due to clarity, not all connections are drawn; dashed rectangles and arrows indicate the danger signal generation via e.g. the necrotic pathway. Different branching in the mutation trees does not influence the fundamental behavior of the system (according Scheidegger et al. [8], modified).

There, $k_{Tik} \cdot T_{ik}$ is the reproduction rate of the tumor sub-population ik (the tumor sub-clones are assumed to form mutation tree with branches k; $k_{T11} \cdot T_{11}$ denotes the reproduction rate of the population $i = 1$ and $k = 1$, for the host tissue, the corresponding rate is $k_{aH} \cdot H$); $k_{eT} \cdot T_{ik}$ represents a spontaneous rate of cell elimination ($k_{eH} \cdot H$ for host tissue); the immune-system – related elimination rate is calculated by $r_{ik}k_{IT} \cdot T_{ik}$ with an interaction coefficient k_{IT}, r_{ik} defines the match with antigen-receptor binding sites [6]. For host tissue, a different coefficient k_{IH} is used; $k_{mut} \cdot q_{il}T_{lk}$ gives the rate of mutation (q_{il} is a matrix representing the topology of the population network, see [8]). Competition between the different tumor sub-populations is included by $k_{TT}T \cdot T_{ik}$ (with the total amount of tumor cells T) and for host tissue by $k_{TH}T \cdot H$; $k_{bH} \cdot H^2$ represents the self-inhibition of host tissue growth. For radiation – induced cell killing, a dynamic linear-quadratic model with a transient biological dose equivalent Γ [10] is used. The radiation-induced death rate is dependent on the radiation dose rate R, the radio-sensitivity coefficients α_H and β_H for host and α_T and β_T for tumor cells. The transient biological dose equivalent Γ is rising with the dose rate R and decaying with a repair constant γ:

$$\frac{d\Gamma_{T,H,I}}{dt} = R - \gamma_{T,H,I}\Gamma_{T,H,I} \tag{2}$$

The indices are indicating that – depending on the cellular repair capability – different repair rate constants $\gamma_{T,H,I}$ have to be applied for tumor cells, host tissue and immune cells or antibodies (for the latter it is assumed to undergo no repair, but the exchange of these cells or antibodies in the tumor compartment lead to a certain "repair" effect which is depending on immigration speed of these cells or antibodies).

The different cell death processes will lead to apoptotic and necrotic cells. Assuming that necrotic cells will contribute mostly to the danger signal, the calculation of this signal is based on the amount of necrotic cells. These cells are "transformed" pre-necrotic tumor cells $N_{p,ik}$ and pre-necrotic or pre-apoptotic host tissue cells $N_{p,H}$, which is related to the system in Eq. 1 (only host tissue cells are considered to be apoptotic by the rate $k_{ap}N_{p,H}$):

$$\frac{dN_{p,11}}{dt} = (k_{eT} + r_{11}k_{IT} + (\alpha_T + 2\beta_T \Gamma) \cdot R) \cdot T_{11} - k_{pn}N_{p,11}$$

$$\frac{dN_{p,ik}}{dt} = (k_{eT} + r_{ik}k_{IT} + (\alpha_T + 2\beta_T \Gamma) \cdot R) \cdot T_{ik} - k_{pn}N_{p,ik} \tag{3}$$

$$\frac{dN_{p,H}}{dt} = (k_{eH} + r_H k_{IH} + (\alpha_H + 2\beta_H \Gamma) \cdot R) \cdot H - (k_{pn} + k_{ap}) \cdot N_{p,H}$$

The pre-necrotic or pre-apoptotic cells subsequently are transformed to necrotic or apoptotic cells at the rate $k_{pn}N_{n,ik}$ and $(k_{pn} + k_{ap}) \cdot N_{p,H}$. The necrotic tumor and host tissue cells are calculated by:

$$\frac{dN_{11}}{dt} = k_{pn}N_{p,11} - k_n N_{11}; \quad \frac{dN_{ik}}{dt} = k_{pn}N_{p,ik} - k_n N_{ik}; \quad \frac{dN_H}{dt} = k_{pn}N_{p,H} - k_n N_H \tag{4}$$

In contrast to the model presented by Scheidegger et al. [8], the danger signal generation includes a two-step process starting with lethally damaged cells which subsequently transforms to "immune-system-activating" cells. For calculating the danger signal, a sigmoidal relationship between the signal strength and the amount of dying cells is assumed:

$$D = \frac{\left[\sum_{i,k} N_{ik} + N_H \right]^2}{L_{act}^2 + \left[\sum_{i,k} N_{ik} + N_H \right]^2} \tag{5}$$

L_{act} governs the steepness of this sigmoidal relation between the amount of necrotic cells and the danger signal D.

The task of the adaptive immune system is the detection of antigen patterns and a response generation based on the presence of the danger signal D. To mimic this process, Scheidegger et al. [8] proposed to model the immune system's adaptability and ability to learn with a perceptron, along with molecular danger signals and antigen-receptor interactions. For this, an antigen pattern vector $\vec{X} = X_i$ can be defined. Every cell of a specific population (tumor sub-clones and host tissue) bears a corresponding pattern, which is defined by the elements of the antigen pattern vector. The presence of a component of the pattern vector is considered to be dependent on the amount of cells bearing this specific component. According to the pattern used in this study (see Fig. 2), the antigen signal strength of the first component for example is given by:

$$X_1 = \frac{\left(\tilde{T}_{11} + \tilde{T}_{12} + \tilde{T}_{13} + \tilde{T}_{14} \right)^2}{(X_{act})^2 + \left(\tilde{T}_{11} + \tilde{T}_{12} + \tilde{T}_{13} + \tilde{T}_{14} \right)^2} \tag{6}$$

with $\tilde{T}_{ik} = T_{ik} + \eta N_{p,ik} + \chi N_{ik}$: pre-necrotic and necrotic cells are considered to contribute to the presence of antigens, but with the weighting factors η and χ. Similar to the sigmoidal relation in Eq. 5, X_{act} influences the activation response.

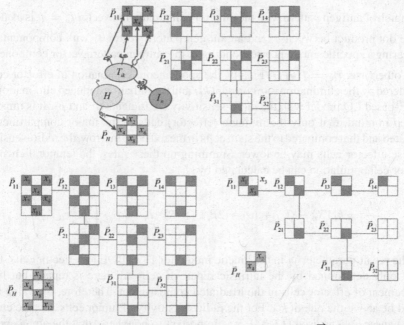

Fig. 2. Mutation tree and population-associated antigen pattern vectors $\vec{P}_{ik}$: Vector components represents epitopes on a specific complex protein or membrane proteins. The number of components is reduced by mutations representing an immunogenic escape. In this study, only an orthogonal setting was investigated. Non-orthogonal settings have been investigated by Scheidegger et al. [8] and they exhibited no fundamental differences in the behavior.

Depending on the presence of a specific antigen signal, the perceptron is used to adapt the corresponding antigen weights w_i for generating the perceptron response by comparing the actual danger signal strength D with the perceptron response Y:

$$\frac{dw_i}{dt} = a \cdot (D - Y) \cdot X_i \tag{7}$$

with the perceptron response Y:

$$Y = \frac{\Sigma^\xi}{Y_{act}^\xi + \Sigma^\xi} \tag{8}$$

where Σ denotes the sum of the antigen signals:

$$\Sigma = \sum_{i=1}^{9} w_i X_i \tag{9}$$

The perceptron response Y directly governs the production of effector cells by the production rate $k_I Y X_n$. The receptor binding of an effector cell of the population I_n with a tumor cell bearing the corresponding antigen will contribute to the tumor cell elimination.

The match of antigen pattern with the effector cell population vector $I_n = \vec{I}$ is evaluated by the dot product between $\vec{I}$ and an antigen pattern vector $\vec{P}$ with components $= 1$ for bearing a specific antigen corresponding to the antigen pattern vector component X_n and 0 otherwise: $r_{ik} = \vec{I} \bullet \vec{P}_{ik}$. Finally, the spontaneous elimination of effector cells is considered by the elimination rate constant k_{eI} and the radiation-induced elimination by a TBDE-based LQ model with the radio-sensitivity coefficients α_I and β_I. It is important to keep in mind, that only the immune (effector) cells in the tumor compartment are irradiated and that compared to the stem cells in the red bone marrow, the radio-sensitivity of these effector cells may be lower. Summing up these rates, the temporal change of effector cell population can be calculated by:

$$\frac{dI_n}{dt} = k_I Y X_n - (k_{eI} + (\alpha_I + 2\beta_I \Gamma_I) \cdot R) \cdot I_n - k_{IT} \cdot \left(\sum_{i,k} r_{ik} T_{ik} \right)_n \tag{10}$$

The repair parameter γ_I in the kinetic model for Γ_I (TBDE for effector cells, Eq. 2) is not only determined by the intrinsic repair of cells (if there is repair) but by the replacement of effector cells in the irradiated compartment. Therefore, the value for γ_I should be above the one of k_{eI}. For the radio-sensitivity of tumor cells, a value close to colon cancer lines is used [13, 14]. It is important to note here, that the alpha and beta values cannot directly compared with the standard LQ model since the kinetic model for the TBDE will reduce cell killing by repair. The effective (corresponding) alpha and beta values are therefore lower in this model (with $\gamma_T = 3d^{-1}$: $\alpha_{T,eff} = 0.128 Gy^{-1}$ and $\beta_{T,eff} = 0.020 Gy^{-1}$), representing more radio-resistant tumor cells such as e.g. cervix carcinoma cells.

The tumor and host tissue growth parameters have been selected based on the following criteria: The tumor is considered as a fast-growing tumor (doubling time of 20 days for all tumor sub-populations; $k_{Tik} = 3.46 \times 10^{-2} d^{-1}$), whereas the host tissue is assumed to repopulate slightly slower. The equilibrium level H_{eq} for host tissue (homeostasis) is set by the values of k_{aH} and k_{eH} to 250 (2.50×10^{11} cells). Assuming an average volume of $2 \times 10^3 \mu m^3$ per cell, the initial compartmental volume is 500 cm³ for 250×10^9 cells. The equilibrium levels for host (H_{eq}) and tumor (T_{eq}) cell population can be calculated by the equilibrium conditions form Eq. 1:

$$T_{eq} = \frac{k_{Tik} - k_{eT}}{k_{TT}} \text{ and } H_{eq} = \frac{k_{aH} - k_{eH}}{k_{bH}} \tag{11}$$

The equilibrium level for the tumor cell population without immunogenic elimination is set to 306 (3.06×10^{11} cells): This corresponds to a scenario where the tumor has less growth limitation than the host tissue.

In this study, 4, 9 and 12 antigen pattern components and 9 corresponding tumor sub-clones are used. The number of sub-clones does not influence the principal dynamics. The structure of the mutation tree is identical to Scheidegger et al. [8] and is displayed in Fig. 2. The used parameter values are summarized in Table 1.

The RT fractionation scheme corresponds to a standard clinical protocol for radical radiotherapy of bladder cancer [15] with 32 fractions of 2 Gy (5×2 Gy per week), leading to a total radiation dose of 64 Gy. Table 1 summarizes the parameter values used for investigation of the presented scenarios. The evolution of the tumor-host ecosystem is simulated over 570 days (before RT), the total simulation time is 1800 days. For numerical integration, a Runge-Kutta algorithm with a time increment of $dt = 10^{-3}$ d is used. Initial conditions are: $H(0) = 250$ (250×10^9 cells), $T_{11}(0) = 10^{-3}$ (10^6 cells) and 0 for all other populations.

3   Results

In the following, the resulting course of the populations (Fig. 3), the perceptron weights (Fig. 4) and effector cells (Fig. 5 & 6) are presented. The development of host tissue and tumor populations (see Fig. 3) exhibit two phases of growth: A first phase before RT and second one after RT (tumor recurrence). In case of 4 and 9 antigen pattern vector components, the main reduction of tumor cells at the end of the first phase coincidences with the application of the RT fractions during the first treatment week. In contrast to this, the case with 12 components exhibits an effector cell - mediated cell killing (strong immune response) before start of RT.

Regarding the perceptron weights (Fig. 4), the values show a fast rise during first week of RT (for 4 and 9 components) or in case of 12 components, before start of RT. The highest values are achieved in the case with 4 antigen pattern vector components, whereas the lowest value can be observed in case of 12 components. After RT, the weights are fixed (frozen) for a certain period at a level dependent to the amount of necrotic cells produced during RT or – in case of 12 components – during first immune response (subsequent RT leads only to minor changes due to the reduced tumor size). As soon as host tissue and tumor cell repopulation reach a certain level (env. 10^9 cells), the weights start to evolve according the dynamics of the re-growing populations.

Table 1. Model parameters (k-parameters normalized to 10^9 cells).

Parameter/Unit	Description	Default value
$k_{T11} = k_{Tik}/\,\mathrm{d}^{-1}$	Tumor growth rate constant	3.46×10^{-2}
$k_{mut}/\,\mathrm{d}^{-1}$	Mutation rate constant	10^{-3}
$k_{eT}/\,\mathrm{d}^{-1}$	Tumor cell elimination rate constant	4×10^{-3}
$k_{TT}/\,\mathrm{d}^{-1}$	Tumor cell growth inhibition	10^{-4}
$k_{IT}/\,\mathrm{d}^{-1}$	Immunogenic tumor cell elimination	1
$k_{HT}/\,\mathrm{d}^{-1}$	Host - tumor cells interaction	10^{-5}
$k_{TH}/\,\mathrm{d}^{-1}$	Tumor - host cells interaction	2.2×10^{-4}
$k_{aH}/\,\mathrm{d}^{-1}$	Host cell growth	3×10^{-2}
$k_{bH}/\,\mathrm{d}^{-1}$	Host cell growth inhibition	1.2×10^{-4}
$k_{eH}/\,\mathrm{d}^{-1}$	Host cell elimination	10^{-5}

(continued)

Table 1. (*continued*)

Parameter/Unit	Description	Default value
$k_{pn}/\,d^{-1}$	Necrotic transformation rate constant	0.5
$k_n/\,d^{-1}$	Necrotic cell elimination	5
$k_{ap}/\,d^{-1}$	Apoptosis rate constant	2
$k_{IH}/\,d^{-1}$	Immunogenic host cell elimination	1
$k_I/\,d^{-1}$	Immune cell production / migration	10
$k_{eI}/\,d^{-1}$	Immune cell elimination	1
Y_{act}	Danger signal activation level	3
ξ	Power of response function	9
X_{act}	Pattern recognition level	2
η	Weight for pre-necrotic cells	0.5
χ	Weight for necrotic cells	0.2
L_{act}	Danger signal param. (Eq. 7)	3
$a\,/\,d^{-1}$	Perceptron learning rate	5
$\alpha_T/\,Gy^{-1}$	Radiation sensitivity tumor cells	0.28
$\beta_T/\,Gy^{-2}$	Radiation sensitivity tumor cells	0.05
$\alpha_H/\,Gy^{-1}$	Radiation sensitivity host tissue	0.05
$\beta_H/\,Gy^{-2}$	Radiation sensitivity host tissue	0.01
$\alpha_I/\,Gy^{-1}$	Radiation sensitivity effector cells	0.1
$\beta_I/\,Gy^{-2}$	Radiation sensitivity effector cells	0.01
$\gamma_T/\,d^{-1}$	Repair constant for tumour cells	3
$\gamma_H/\,d^{-1}$	Repair constant for host tissue	10
$\gamma_I/\,d^{-1}$	Repair constant for immune cells	2
$R\,/\,Gy/min$	Radiation dose rate	0.14

In all presented scenarios, the perceptron weights for host tissue evolve into negative values during the re-growth of the tumor. Regarding Eq. 8 and Eq. 9, these negative weights reduce the perceptron response and therefore have an immune - suppressive effect. Corresponding to this, a second effector cell – mediated response in case of 12 antigen pattern components is missing, the population sizes develop similar to the case with 9 components. This effect is also visible in Fig. 5, where the number of effector cells is displayed. In all cases, only weak effector cell recruitment takes place during the second phase. Beside the anti-tumor response, an anti-host response can be observed at the end of every phase. In case of 4 antigen pattern vector components, the anti-host immune response during RT is stronger than the anti-tumor response.

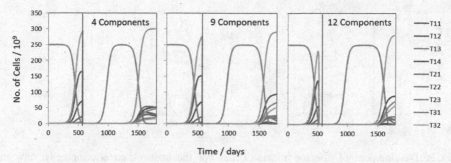

Fig. 3. Development of the host - and different tumor – cell populations: The vertical red line indicates the time point of RT start (day 570). In contrast to the scenarios with 4 or 9 antigen pattern vector components, the case with 12 components exhibits an effector cell – mediated elimination prior to RT. After RT, the tumor starts to regrowth and approaches in every scenario the equilibrium level of 306×10^9 cells (Color figure online).

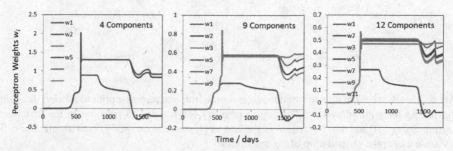

Fig. 4. Evolution of the perceptron weights: Prior to RT, all weights rise due to necrotic cells (mainly tumor cells). The presence of host tissue cells in this stage leads to an anti-host immune reaction. RT as well as the prior immune response (in case of 12 components) lead to a separation of the tumor- and host associated weights, which becomes more pronounced during host tissue regrowth (due to lack of a danger signal). In every scenario, the host-associated weights (only w2 is displayed, the other host-associated weights are identical) evolve into negative values.

In Fig. 6, a section of the effector cell response during the first immune-mediates response for the case with 12 components and during RT for the case with 9 components is shown. In case of RT-induced response, the radiation-mediated cell killing is visible as spikes at the position of each RT fraction.

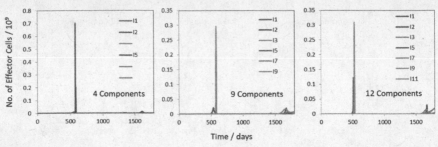

Fig. 5. Development of effector immune cells: For the host-related effector cells, only the population I2 (red line) is displayed, the other populations behave identical (Color figure online).

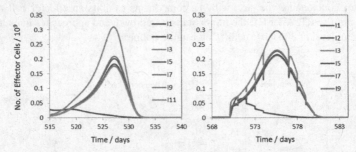

Fig. 6. Development of effector immune cells during immune response prior to RT (left diagram) and during RT (right diagram): The impact of every RT fraction (5 fractions in the first week starting at day 570 and 4 of 5 fractions of the 2nd week starting at day 577) onto the effector cells is visible as a spike-shaped drop of the cell number.

4   Discussion and Conclusions

The model is not intended to be applied directly to real patients. At this stage, most of the parameter values are not based on biological data. It would be interesting to develop novel strategies for assessing this information by studies in vitro and in vivo. Regarding clinical trials, patient specific variations may not allow to define a "standard" set for such parameters. In this regard, the proposed model may serve as a lab in silico for evaluating different scenarios of patient-specific immune responses.

In the case of 12 antigen vector components, a strong immune response is triggered prior to RT. However, during tumor recurrence, the immune response remains weak for all scenarios (4, 9 and 12 components). Similar to Scheidegger et al. [8], the appearance of negative host weights can be observed after RT. These negative weights have an immune-suppressive effect. This phenomenon after RT seems to be a general behavior of this system. Under certain conditions (especially during RT in the case of four antigen pattern vector components), the presence of host-associated antigen patterns in combination with a danger signal triggers an anti-host immune reaction, which can be interpreted as an inflammatory process including host tissue. This may have tumor suppressive as well as tumor promoting effects, depending on chronic versus acute inflammatory scenarios

[2, 16]. An increased number of antigen vector components seems to lead to a better distinction between host tissue and tumor cells.

No vaccination effect by RT was observed in the investigated scenarios. This may be explained by the fact that there is no immunogenic memory implemented and the perceptron is trained by a reinforced learning process. In addition, only a tumor - host compartment is considered. In case of a real patient, host tissue is present at locations where a danger signal is missing or not generated. Assuming that the adaptive immune system does not only process information locally, a systemic immune response has to integrate perceptron responses from different compartments. The following expansions of the model would be needed to investigate vaccination effects: (1) Implementation of immunogenic memory by memory cell populations; (2) at least one or two additional compartments for rest body (host only) and metastasis to investigate eradication of distant, non-irradiated metastases or so-called abscopal effects [17, 18]; (3) multi-compartmental information processing with local perceptrons which exchange information (may be considered as a multi-layer neuronal network).

Albeit only a small number of scenarios have been investigated in this study, some important conclusions are already possible: To understand anti-tumor immune response, the whole ecosystem dynamics should be regarded. To compare these theoretical results with the real patients, time-resolved data regarding type and amount of immune cells infiltrating the tumor compartment are needed. This would strongly support further advancements of the proposed framework using a perceptron-based approach.

Acknowledgements. This project (Hyperboost; www.Hyperboost-h2020.eu) has received funding from the European Union's Horizon 2020 research and innovation programme under the Marie Skłodowska-Curie grant agreement No 955625.

References

1. Alfonso, J.C.L., Papaxenopoulou, L.A., Mascheroni, P., Meyer-Hermann, P., Hatzikirou, H.: On the immunological consequences of conventionally fractionated radiotherapy. iScience **23**, 100897 (2020). https://doi.org/10.1016/j.isci.2020.100897
2. Di Maggio, F., et al.: Portrait of inflammatory response to ionizing radiation treatment. Journal of Inflammation **12**(14), 111 (2015) https://doi.org/10.1186/s12950-015-0058-3
3. Frey, B., Rückert, M., Deloch, L., Rühle, P.F., Derer, A., Fietkau, R., Gaipl, U.S: Immunomodulation by ionizing radiation-impact for design of radio-immunotherapies and for treatment of inflammatory diseases. Immunol. Rev. **280**(1), 231–248 (2017)
4. Pienta, K.J., McGregor, N., Axelrod, R., Axelrod, D.E.: Ecological therapy for cancer: defining tumors using an ecosystem paradigm suggests new opportunities for novel cancer treatments. Translational Oncology **1**, 158–164 (2008)
5. Basanta, D., Anderson, A.R.A.: Exploiting ecological principles to better understand cancer progression and treatment. Interface Focus **3**, 20130020 (2015)
6. Merlo, L.M.F., Pepper, J.W., Reid, B.J., Maley, C.C.: Cancer as an evolutionary and ecological process. Nat. Rev. Cancer **6**, 924–935 (2006)
7. Enderling, H., Wolkenhauer, O.: Are all models wrong? Comput Syst Oncol. **1**(1), e1008 (2020)

8. Scheidegger, S., Mikos, A., Fellermann, H.: Modelling artificial immune – tumor ecosystem interaction during radiation therapy using a perceptron – based antigen pattern recognition. ALIFE 2020: The 2020 Conference on Artificial Life July 2020, The MIT Press Journals, pp. 541–548 (2020)

9. Eftimie, R., Bramson, J.L., Earn, D.J.D.: Interactions between the immune system and cancer: a brief review of non-spatial mathematical models. Bull. Math. Biol. **73**(1), 2–32 (2011). https://doi.org/10.1007/s11538-010-9526-3

10. Rosenblatt, F.: The perceptron: A probabilistic model for information storage and organization in the brain. Psychological Review **65**(6), 386 (1958). https://doi.org/10.1037/h0042519

11. Matzinger, P.: The danger model: a renewed sense of self. Science **296**(5566), 301–305 (2002)

12. Scheidegger, S., Lutters, G., Bodis, S.: A LQ-based kinetic model formulation for exploring dynamics of treatment response of tumours in patients. Z. Med. Phys. **21**, 164–173 (2011)

13. Leith, J.T., Padeld, G., Faulkner, L.E., Quinn, P., Michelson, S.: Effects of feeder cells on the x-ray sensitivity of human colon cancer cells. Radiother. Oncol. **21**(1), 53–59 (1991)

14. Van Leeuwen, C.M., et al.: The alpha and beta of tumours: a review of parameters of the linear-quadratic model derived from clinical radiotherapy studies. Radiat. Oncol. **13**, 96 (2018)

15. The Royal College of Radiologists: Radiotherapy dose fractionation, third edition. 2019, The Royal College of Radiologists, 63 Lincoln's Inn Fields, London WC2A 3JW, United Kingdom. https://www.rcr.ac.uk/system/files/publication/field_publication_files/bfco193_radiotherapy_dose_fractionation_third-edition-bladder_0.pdf. Accessed 21 June 2021

16. Grivennikov, S.I., Greten, F.R., Karin, M.: Immunity, inflammation, and cancer. Cell **140**, 883–899 (2010)

17. Formenti, S.C., et al.: Radiotherapy induces responses of lung cancer to CTLA-4 blockade. Nature Medicine **24**, 1845–1851 (2018)

18. Golden, E.B., et al.: An abscopal response to radiation and ipilimumab in a patient with metastatic non-small cell lung cancer. Cancer Immunol. Res. **1**, 365–372 (2013)

Two-Level Detection of Dynamic Organization in Cancer Evolution Models

Gianluca D'Addese[1] , Alex Graudenzi[2,3] , Luca La Rocca[1] ,
and Marco Villani[1,4(✉)]

[1] Department of Physics, Informatics and Mathematics,
University of Modena and Reggio Emilia, Modena, Italy
{luca.larocca,marco.villani}@unimore.it

[2] Institute of Molecular Bioimaging and Physiology, Italian National Research Council
(IBFM-CNR), Segrate, Milan, Italy
alex.graudenzi@unimib.it

[3] Bicocca Bioinformatics, Biostatistics and Bioimaging Centre (B4),
Milano-Bicocca University, Milan, Italy

[4] European Centre for Living Technology, Venice, Italy

Abstract. Many systems in nature, society and technology are composed of numerous nonlinearly interacting parts. The dynamic organization of these systems often allows the emergence of intermediate structures that once formed deeply affect the system, and therefore play a key role in understanding its behavior. An interesting hypothesis is that the simultaneous analysis of a system at both levels of description (the microlevel of the relationships between single entities, and the mesolevel constituted by their dynamically organized groups) can allow a better understanding of the phenomenon under examination. In this work we apply this idea to a cancer evolution model, of which each individual patient represents a particular instance. Specifically, in order to validate the idea we analyze the same synthetic dataset – whose ground truth is known – with two methods of analysis, and we merge the results in an innovative way. In doing this, we also evaluate the effectiveness of a new method of reconstructing networks of relationships.

Keywords: Complex systems analysis · Information theory · Relevance index · Cancer evolution

1 Introduction

Many systems in nature, society and technology are composed of numerous parts that interact in non-linear ways [1, 2]. In these systems the emergence of intermediate structures is frequently observed [3, 4]. Paradigmatic examples are present in biology, where for instance we can identify organs, which are necessary to describe a multicellular organism (the *macrolevel*) composed by a huge number of cells (the *microlevel*). In the same way organelles can be regarded as intermediate entities between macromolecules and cells, and tissues can be regarded as intermediate between cells and bodies.

© The Author(s), under exclusive license to Springer Nature Switzerland AG 2022
J. J. Schneider et al. (Eds.): WIVACE 2021, CCIS 1722, pp. 207–224, 2022.
https://doi.org/10.1007/978-3-031-23929-8_20

Similar organizational aspects (the presence of structures at an intermediate level) can be found in different kinds of systems. In social systems we can observe several intermediate bodies between the state and individuals: parties, associations, movements, trade unions, etc. At a bigger scale, alliances, federations and leagues of nations are present, intermediate organizations between the states and the whole of mankind. Likewise, we can observe the emergence of technological organizations based on the interaction between computers in artificial systems, or the presence of dynamic structures composed by computers, (semi)automatic systems and human beings in socio-technological systems. Intermediate-level structures, once formed, deeply affect the system as a whole, and therefore play a key role in understanding its behavior.

The detection of intermediate structures in complex systems is not always a trivial task, while in contrast their characterization can lead to a meaningful description of the overall properties of the system, and in this way to its understanding [2–4]. A large part of these structures is characterized by groups of variables (genes, chemical species, individuals, agents, ...) that appear to be well coordinated among themselves and have a relatively weaker interaction with the remainder of the system. This general observation is the basis of several algorithms aiming to identify intermediate level structures in different fields: notable examples are the identification of functional neuronal regions in the brain [5, 6], autocatalytic systems in chemistry [7–9], the identification of communities in socio-economic systems [10, 11] and the detection of specific groups of genes governing the dynamics of a genetic network [8, 9].

In a recent paper, some of us observed that variants of these techniques can allow the reconstruction of pairwise relationships between variables in gene knock-out experiments [12] - a result comparable with that of other existing approaches [13, 14]. As it is, the composition of these relationships can lead to the construction of a network of binary interactions, which constitutes a possible representation of the dynamic organization of the system under examination. An interesting idea is that the simultaneous application of the two approaches on the same dynamic system, resulting in a collection of dynamic groups together with a graph of binary relationships, could allow a better understanding of the system [12].

In this work we want to test the validity of the above idea by applying it to a particularly relevant case, in which we look for the dynamic organization responsible for the set of genomic alterations that are observed in cancer patients. In human cancer each patient represents an instance of the evolution process of that specific cancer type: so, we can consider each patient as a single observation (a dynamic state) of the system under examination (the evolution of that cancer). The juxtaposition of these observations could allow the extraction of information about the dynamic process that produced them. In particular, we are interested in identifying relationships between genome mutations that lead to cancer progression, in order to recognize the presence of distinct cancer progression patterns. A successful validation of the approach could allow the creation of new tools for the analysis of (not only) biomedical data.

In this paper we combine the reconstruction of a network of relationships between pairs of mutations with the search for dynamically organized groups of mutations. We validate this approach by analyzing data regarding synthetic patients, simulated from a model of cancer progression, and we compare our results with those of some state-of-the-art algorithms. We conclude that the simultaneous identification of mesolevel structures and pairwise relationships can improve the reconstruction and understanding of the underlying dynamic organization.

In Sect. 2 we present the RI method (based on indices derived from information theory), a recent method for the reconstruction of relationship networks based on it and the algorithms with which we intend to compare the latter. Section 2 presents also the two-level analysis strategy we intend to use. Section 3 discusses the case of cancer progression and the model we used to produce the artificial data. Section 4 presents the analyses performed and their results. The results in case of sequencing noise and relative scarcity of data are also discussed. Finally, Sect. 5 presents the conclusions and some indications for future work.

2 The Identification Methods

2.1 Introduction

The present work is mainly based on the Relevance Index (RI) metrics [8, 9, 12, 15–17]: a set of information-theoretical metrics[1] for the analysis of complex systems that can be used to detect the principal interacting structures (Relevant Sets, or RSs in the following) within them, starting from the observation of the status of system variables over time (or from observations not necessarily exhibiting a temporal order) [16, 18–20]. The method can be applied by using different indices: in this paper we use zI index, as described below.

In Sect. 2.2, we briefly review our method: the computation of the zI index, and the "iterative sieving" procedure we use to group variables based on the index values. The result of this approach is a list of groups, composed of variables (individuals, agents,...) dynamically connected to each other. In the final part of Sect. 2.2 we concisely present the method published in [17], through which it is possible to identify a set of binary relationships - that is, a graph – that constitute a possible representation of the network of relationships in the system under examination. A more detailed exposition can be found in [12]. In Sect. 2.3 we present the combination of the two methods - that is, the idea we want to test in this paper. In Sect. 2.4 we present a useful original comment on the method itself. Finally, in Sect. 2.5 we present two well-known methods aiming to reconstruct networks of binary relationships with which we will confront.

2.2 The RI Method

Considering a system composed of m random variables $X_1, X_2, ...,X_m$ we suppose that S_k is a subset composed of k elements, with $0 < k < m$. Our purpose is to identify subsets of variables that behave in a somehow coordinated way, i.e., the variables belonging to the subset are integrated with each other, much more than with the other variables of the system. As these subsets can be used to describe the whole system organization, they are named Relevant Subsets (RSs).

---

[1] See Cover and Thomas [25] for a proper introduction to information theory.

In order to find these structures, we observe the behavior of the system under examination and measure the level of internal coordination of all its subsystems. In the case of pairs of variables, this measure can be declined as mutual information, while in the case of groups composed of $k > 2$ variables, a generalization called Integration can be used:

$$I(S_k) = \sum_{s \in S_k} H(s) - H(S_k) \tag{1}$$

where $H(S_k)$ denotes the joint entropy of the variables in S_k and $H(s)$ is the marginal entropy of X_s.

It is advisable to account for the size of the subsystem and the cardinality of its state space by using a z-score [16, 17]:

$$zI(S_k) = \frac{2nI(S_k) - \langle 2nI(S_k) \rangle}{\sigma(2nI(S_k))} \tag{2}$$

where n is the number of observations, and $<2nI(Sk)>$ and $\sigma(2nI(S_k))$ are, respectively, the average and the standard deviation of the measurements related to a matching homogeneous system (a system of the same size of the system under examination, whose variables are mutually independent). These averages can be effectively approximated through a Chi Square distribution, whose degrees of freedom depend on the size of the subset and on the cardinality of its alphabet [6, 9, 16, 17]. Interestingly, the integration is related to the identification of dynamical criticality in complex systems [19, 21].

The list of candidate RSs can be very long, with many partial overlaps. To identify the truly essential subsets a sieving algorithm is hence implemented, by preserving sets that are not included and do not include any other set that has a higher zI value [16, 17, 23, 24]. These sets represent the building blocks of the system's dynamic organization. The variables belonging to each subset are merged into a single new variable - hereafter called a "group variable" - which inherits the behavior of the subset (the combination of states of the variables belonging to it). It is possible to iterate the procedure, repeating the analysis and aggregating variables, until the zI index is so low that it can no longer justify further mergers. This approach allows one to identify a plausible organization of the system in terms of non-overlapping groups of variables [17, 22, 23].

The final collection of variables (single variables, or group variables) – called Relevant Sets, or RSs for short - is the result of the method, and it represents the fundamental dynamic organization of the system. We will refer to this approach with the name of *zI_full* method. In the following we will judge as not statistically relevant the groups showing a zI value below a threshold, whose reference values *zI_theta* = 3.0 (or *zI_theta* = 5.0) derives from statistical considerations and corresponds to a normalized distance of 3 (or 5) standard deviations from the reference condition of variable independence [16, 17].

As anticipated, the RI method was not designed to reconstruct the topology of systems whose organizational structure is supposed to be representable by a graph (the RI method is more related to the reconstruction of hypernetworks [24]). However, we showed in [17] that limiting this approach, without iterations, to just pairs of variables allows the identification of a set of binary relationships, that is, a graph. In this graph it is possible to eliminate a large part of the links whose existence can be derived from other already present links (the "epiphenomenal" links) by using an information theoretic property called Data Processing Inequality (in short, DPI) [25]. In this way we arrive at the identification of an essential network of relationships, the reconstruction of the dynamic network underlying the system - assuming that this organization can be represented by a graph. We will refer to this approach (and when it does not cause confusion, also to the resulting graph) with the name of *zI_graph*.

2.3 Merging Groups and Graphs

As described in Sect. 2.2, it is possible to reconstruct a graph of relationships (each relationship locally connecting only two variables - "micro" level) by using the *zI_graph* method, and at the same time identify the presence of large and integrated groups of variables (the RSs - "meso" level) by using the *zI_full* method (Fig. 1). In this paper we show how the two methods can be usefully combined in the analysis of the same system (the whole reality - "macro" level).

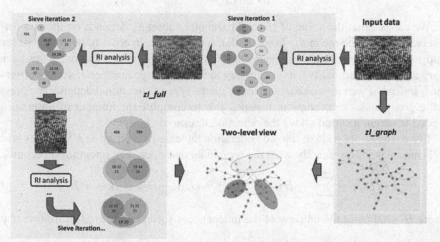

Fig. 1. The diagram shows the two methods presented in the text (*zI_full*, blue flow of actions, and *zI_graph*, pink flow of actions), applied on the same dataset. The combination of the two methods allows a better understanding of the system under examination. In particular, the mesolevel analysis (*zI_full* method) identifies mesolevel groups (the RSs), while the microlevel analysis (*zI_graph* method) elucidates the links between the individual variables. The overlap of the two analyses (two-level view) allows the identification of possible links between different RSs (the links between nodes that also connect two different RSs). On the other hand, the mesolevel analysis limits the connectivity of the underlying network, indicating that some relationships, although present at the microlevel, are not such as to allow the merging of the RSs.

2.4 A Note on the Number of Observations

The calculation of the mean and standard deviation of the homogeneous system (useful in the comparison of candidate RSs with different sizes or alphabets) is very onerous: in case of simulations, it requires the realization and analysis of thousands of instances of the homogeneous system, all having the dimensions of the system under examination. As already commented, this explicit calculation can be avoided by resorting to the observation that the distribution of $2nI$ - for large value of n - follows a Chi Square distribution, whose degrees of freedom depend on the size of the subset and on the cardinality of the alphabet of the variables belonging to it [17]:

$$d_k = \prod_{j=1}^{k} |\chi_j| - 1 + k - \sum_{j=1}^{k} |\chi_j| \tag{3}$$

where $|\chi_j|$ denotes the cardinality of the alphabet of the variable j. The degrees of freedom d_k coincides with the mean of the Chi Square distribution (which does not depend on n), and $\sigma = sqrt(2^* d_k)$ is its standard deviation. We therefore obtain:

$$zI(S_k) = \frac{2nI(S_k) - \langle 2nI(S_h) \rangle}{\sigma(2nI(S_h))} \approx \frac{2nI(S_k) - d_k}{\sqrt{2d_k}} \tag{4}$$

We can see that the value of zI for a particular subset S_k depends on the number of observations and the mean d_k. In particular, we note that, in order to be able to observe groups having a high number of degrees of freedom (large groups and/or groups having a large alphabet cardinality), it is necessary to collect a large number of system observations. Indeed, if we can make a hypothesis on the type of functional dependence present in the group under examination, it is possible to compute the number of observations needed to obtain a signal above the detection threshold zI_theta.

For example, let us take into consideration the case of a group of k variables where $k-1$ variables follow exactly what the remaining one does. The integration becomes:

$$I(S_k) = \sum_{s \in S_k} H(s) - H(S_k) = kH_{k0} - H_{k0} = (k-1)H_{k0} \tag{5}$$

where H_{k0} indicates the entropy of the independent variable. From (3) it follows that:

$$zI(S_k) = \frac{2nI(S_k) - d_k}{\sqrt{2d_k}} = \frac{2n(k-1)H_{k0} - d_k}{\sqrt{2d_k}} \tag{6}$$

It is therefore possible to determine the minimum number of observations needed to exceed the detection threshold (for the case under consideration):

$$\frac{2n(k-1)H_{k0} - d_k}{\sqrt{2d_k}} > zi_{theta} \rightarrow n > \frac{zi_{theta}\sqrt{2d_k} + d_k}{2(k-1)H_{k0}} \tag{7}$$

Note that the particular dependency chosen (a perfect sequence of a leader variable - see also [9]) is one of the strongest dependencies that can be identified. Therefore, the value given by Eq. (7) represents a lower bound on the number of observations needed in practice.

2.5 Reconstructing the Network of Binary Relationships

One of the results of this paper is that of reconstructing the dynamical organization underlying the cancer progression process. We obtain this result in two different ways: by reconstructing the network of binary relationships between the involved variables ("micro" level) and by identifying the dynamically relevant sets ("meso" level). While there is no counterpart to mesolevel analysis, there are several methods in the literature regarding the reconstruction of relationship networks. At the microlevel, in order to evaluate our proposal, we take into consideration a system that uses an approach similar to ours (mutual information measures – ARACNE-AP) and a method specifically designed for the reconstruction of tumor progression trees (CAPRI).

The ARACNE-AP [26] algorithm is an improved version of the ARACNe method (Algorithm for the Reconstruction of Accurate Cellular Networks) in [13], which uses a microarray expression profile to reconstruct a gene regulatory network, taking a pairwise information theoretic approach. This ameliorated version is based on an Adaptive Partitioning strategy for estimating the Mutual Information, which achieves a boost in computational performance, while preserving the network inference accuracy of the original algorithm. The software we use [27] takes as input the microarray expression profile and a transcription factor (TF) list. This last input is used to filter the connections to be displayed (the ones involving nodes in this input list). Since our interest is not restricted to some nodes in particular, but we aim to the reconstruction of the whole structure, our TF list will contain all the nodes of the system.

The CAPRI algorithm [14], which is included in the TRONCO R package for translational oncology [28], relies on Suppes' theory of probabilistic causation [29] and combines maximum likelihood estimation with regularization (e.g., via BIC or AIC) and bootstrap, and extracts general directed acyclic graphs that depict branched, independent, and confluent evolution. CAPRI was proven effective in delivering robust and accurate models from cross-sectional mutational profiles of distinct cancer (sub)types and was included in the PICNIC pipeline for cancer evolution [30]. Recent extensions of the approach have proven effective in stratifying patients in statistically significant risk groups [31].

3 Independent Progressions in Cancer Evolution Models

3.1 The Problem

Cancer is a multi-factorial disease ruled by Darwinian evolution, i.e., by the positive selection of cell subpopulation characterized by the emergence/accumulation of specific (epi)genomic alterations, typically known as drivers [32, 33].

For instance, it is known that certain genomic alterations (single-nucleotide variants, structural variants, copy-number alterations, etc.) hitting specific genes can confer cancer cells a certain selective advantage, which allows them to proliferate in an uncontrolled fashion, escape apoptosis, evade the immune system and diffuse in other organs [34]. Such genomic alterations are typically inherited during cell divisions, so evident accumulation patterns are observed in the history of any given tumor.

Even if the reliable identification of such driver events is one of the grand challenges in cancer research, finding possible regularities across different patients affected by the same tumor (sub)type may be essential to deliver explanatory models of the disease, to dissect tumor heterogeneity and, possibly, to produce reliable predictions. This might in turn drive the development of efficient anticancer therapeutic strategies [35].

The widespread diffusion of next-generation sequencing experiments (e.g., whole-genome, whole-exome, targeted) allows one to effectively detect the presence of genomic alterations in each tumor, for instance by calling variants from bulk samples [36].

Starting from cross-sectional mutational profiles collected from different patients of a given tumor (sub)type, it is then possible to employ state-of-the-art statistical approaches to reconstruct probabilistic graphical models depicting the most likely trends of accumulation of driver events [37, 38].

3.2 Description of the Test Case

We tested the accuracy and robustness of our approach on simulated datasets in different experimental scenarios.

Synthetic binary datasets representing cross-sectional mutational profiles of cancer patients were sampled from a number of distinct generative topologies, similarly to [39]. We randomly generated several forest generative topologies, composed by different constituting trees in which each node represents an arbitrary genomic alteration, whereas each edge corresponds to the subsequent evolutionary step and is characterized by a randomly assigned conditional probability (Fig. 2a). Forest topologies were employed to simulate the case in which different patients of a given dataset belong to distinct cancer subtypes, corresponding to the distinct constituting trees.

More in detail, each binary dataset is an n (patients) $\times$ m (genomic alterations) dataset in which each entry is 1 if an alteration is present, 0 otherwise. Each dataset was randomly sampled starting from the root of the model (in this case, the root is a "fake" node upstream to the roots of the distinct constituting trees). In order to simulate sequencing errors, we included false positives/negatives in the dataset with probability $v \in (0, 1)$: i.e., false positives $(\varepsilon+) = $ false negatives $(\varepsilon -) = v/2$.

It is complicated to estimate the level of false positives and false negatives that might be present in real-world mutational profiles of cancer patients, as this aspect is highly influenced by the adopted sequencing technology and by the possible technical limitations of single experiments, as well as by the various computational steps employed in the variant calling procedure, including QC [40].

However, the range of noise rates used in the generation of our synthetic datasets was chosen in accordance with similar studies on the topic (see, e.g., [14, 31]) and allowed us to assess how the accuracy and robustness of any computational method is progressively influence by increasing amounts of data-specific errors, especially with respect to the noise-free scenario.

We finally scanned the number of samples of the dataset, to assess the impact of samples size on the accuracy and robustness of the inference.

In brief, we generated (Fig. 2c):

- forests with 2 trees - topologies indicated with the acronyms from F2T1 to F2T5

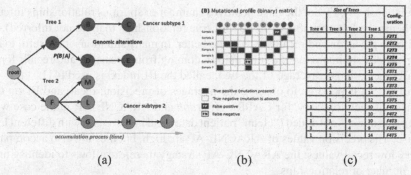

Fig. 2. Synthetic data generation. (a) A graphical representation of the simulated tumor progression tree. The nodes are genomic alterations; presence of edges indicate a randomly assigned conditional probability different from zero. The flow of the edges determines the accumulation paths. Different trees correspond to different tumor progressions/subtypes. (b) The resultant simulation matrix. Different samples are generated to construct several patients. The final result of this task is a binary matrix reporting the presence (black cells) or the absence (white cells) of mutations. A sequence noise can be added (resulting in false negatives and false positives). (c) The number of nodes belonging to each progression tree (for every configuration of this work).

- forests with 3 trees - topologies indicated with the acronyms from F3T1 to F3T5
- forests with 4 trees - topologies indicated with the acronyms from F4T1 to F4T5

All topologies are composed by $m = 20$ nodes. For each topology, we sampled binary datasets with $n = 50, 100, 150, 200$ samples and 3 noise levels: $v = 0, 0.1$, and 0.25.

4 Experimental Results

4.1 Introduction

Recall that have two levels of observation: the microlevel (the list of relationships) and the mesolevel (the list of relevant sets, RSs). In this section we show the results of applying the zI index for the two purposes, also proposing a combination of the two approaches.

4.2 The Network of Relationships

As anticipated, to evaluate the effectiveness of the zI index in the reconstruction of relationship networks (zI_graph method) we decided to compare it with the CAPRI and ARACNE-AP methods. As regards the zI, we use both thresholds $zI\_theta = 3.0$ and $zI\_theta = 5.0$: in the following we therefore use the abbreviations $zI3$ and $zI5$ to distinguish the two situations. To give a quantitative measure of the algorithms' performance we make use of the precision and recall indexes [41], calculated by using the cancer progression trees on which the creation of the simulated patients is based. In our context, precision and recall are defined as follows: precision $=$ TP/(TP + FP) and recall $=$ TP/(TP + FN), where TP are the true positives (number of correctly inferred true relationships), FP are the false positives (number of spurious relationships inferred) and FN are the false negatives (number of true relationships that are not inferred). The closer both precision and recall are to 1, the better. In many situations it is useful to have a single index summarizing the information coming from precision and recall: for this purpose, the harmonic average of the two, called the F1 index, is used [41].

In Fig. 3 it is possible to observe the averages of precision, recall and F1 on the 5 instances of each topology, for CAPRI, zI_graph and ARACNE-AP, in the case where sequencing noise is not added ("clean" patient data), and in two cases with different levels of noise. The precision values of ARACNE-AP are high, but they are also accompanied by very low recall values: the ARACNE-AP strategy therefore allows to identify only a small number of relationships.

The recall values also explain the otherwise bizarre improvement in precision of ARACNE-AP with increased sequencing noise: what actually happens is that fewer and fewer relationships are identified, only those that have really high index values. Indeed, we reach the extreme of systems in which only one relationship is identified (so evident as to be exact, and therefore with a precision value equal to 1.0). In some cases, no relationship is found. The lack of a consistent number of relationships does not allow the correct identification of cancer progression trees: ARACNE-AP is therefore not useful for the purposes of this work and will not be further discussed.

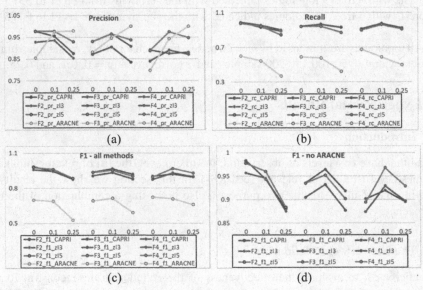

Fig. 3. (a) The averages of precision over the 5 instances of each topology presented in the main text, for CAPRI, $zI3\_graph$, $zI5\_graph$ and ARACNE-AP. The trend of the averages is shown as the noise level varies (0%, 10%, 25%), for each topology involved (F2, F3 and F4 - the three groups evident in the image). (b) The averages of recall. (c) The averages of the F1 index. (d) The averages of the F1 index for CAPRI, zI_graph with threshold 3 and zI_graph with threshold 5 only (ARACNE-AP is not present).

CAPRI, $z13$ and $z15$ instead show very similar recall values, even when the noise intensity varies (Fig. 3b). This fact means that the synthesis provided by the F1 index is able to adequately summarize both precision and recall (Fig. 3c): in the rest of the paper we therefore make use of this index. The F1 index shows that these three algorithms are substantially equivalent (Fig. 3d), with $z15$ showing a slightly higher performance than the other two (Fig. 3a and Fig. 3b).

4.3   Identification of Branches and Trees

In order to correctly recognize the cancer progression process, the identification of a set of nearly correct links is not enough: it is necessary to reconstruct the links' dynamical organization. In the cases under examination, the process is organized in two or more separate structures (the cancer progression trees): the structure of the reconstructed links does not necessarily match this property.

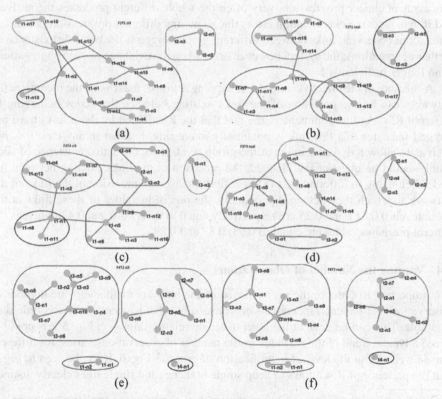

Fig. 4. Results of the analysis of the topologies F2T1 (points a, b), F3T4 (points c, d) and F4T2 (points e, f). In the left column (points a, c, e) the mapping of the identified RSs on the network reconstructed by $z1\_graph$; in the right column (points b, d, e) the mapping of the identified RSs on the real network. The RSs are always contained within the same cancer progression tree (in figure trees $t1$, $t2$, $t3$ and $t4$), and if they do not fully identify it, they identify at least the main branches.

The *zI_full* methodology has the explicit objective of identifying groups (a level higher than that of the single links): in Fig. 4 we can observe their location within the reconstructed network (*zI-graph*) and within the "real" network. We can observe that:

- often the groups coincide with the single trees.
- when this is not the case

  o groups always identify nodes within the same tree (there is no confusion between nodes belonging to different cancer progression trees[2])
  p the groups coincide with (part of the) single branches of the tree[3]

The presence of the groups therefore allows to avoid mixing two different cancer progressions into the same process, as it might happen by identifying such progressions as the mere presence of simply connected components in the reconstructed graph – Fig. 4. Indeed, the *zI_full* methodology is correctly identifying the branches of the different processes of cancer progression, very often the whole different processes themselves, and it avoids severe classification errors due to mixing different progression processes. The fact of observing links between different groups suggests the possible existence of further aggregations: the system however stops when the evidence of such aggregations is no longer sufficiently high.

A measure of the effectiveness of the mapping action of the RSs on the reconstructed networks can be the precision of the links existing between the nodes belonging to different RSs. As just commented, the fact that the RSs joined by these links have not merged indicates that this link is potentially interesting, but that in any case it is not such as to allow a dynamic union at the group level. If so, the actual existence of such a link should be questionable. Indeed, the precision of the links existing between the nodes belonging to different RSs is lower than the average precision of the links of the network to which they belong. Specifically, the precision values of these links at the noise levels 0.0, 0.1 and 0.25 are respectively equal to 0.70, 0.46 and 0.47, whereas the general precision values are equal to 0.93, 0.87 and 0.84.

4.4 Varying the Number of Observations

It is important to observe the performance of the analysis methods as the number of observations (of patients) increases - or, conversely, what happens if there is little data available. This can be done at the level of single relationships – in Fig. 5 it is possible to see a typical trend of the F1 index as the number of observations varies, for different noise levels - or at the level of identification of groups - Fig. 6. It can be seen in Fig. 6 that the presence of RSs allows to keep single branches and single trees clearly distinct.

---

[2] In our experiments there are only two exceptions, both consisting of a small tree composed by a single leaf. The noise present in the limited sample of observations provides sufficient evidence for such aggregation; in general, for any method of analysis, it is always possible that a "bad" set of observations will bring incorrect evidence. No cases of incorrect mixing of large trees (or even just large branches) have been recorded.

[3] Confusions between branches of the same tree are rare, and they occur only in presence of high noise levels.

As already commented, large dynamic groups organized in a non-obvious way typically express a large number of different states. Having few observations available, it is therefore very unlikely to be able to identify their presence. Vice versa, an increasing number of observations can allows to identify of larger and larger dynamic groups (if any): this phenomenon is highlighted by Fig. 7, which shows the average and maximum values of the RSs identified in the various topologies, as the number of observations increases. This fact explains the enlargement and mergers of the identified RSs of the analysis series in Fig. 6.

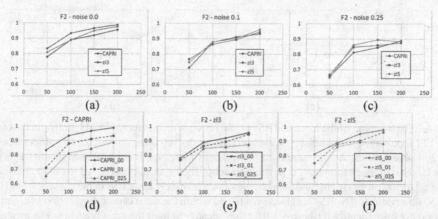

(a) (b) (c)

(d) (e) (f)

Fig. 5. First row: average F1 indexes on the F2T1 topology, as the number of observations (patients) varies, with noise level 0% (a), 10% (b) and 25% (c). Second row: (d) Average F1 indexes of CAPRI on the F2T1 topology as the number of observations varies, all noise levels. (e) Average F1 indexes of *zI3* on the F2T1 topology as the number of observations varies, all noise levels. (f) Average F1 indexes of *zI5* on the F2T1 topology as the number of observations varies, all noise levels. We can observe that the zI index is more robust than CAPRI with respect to the introduction of sequencing noise.

As a final consideration, we can note that the average and maximum sizes of the groups identified in the various topologies grows as the number of observations (patients) increases (Fig. 7). Note that Eq. (7), in case of binary variables and Hko = 1 (a very changing leader variable) indicates the need for n > ~35 observations to identify dynamic groups composed of at least 9 variables (n > ~194 for 12 variables).

As already commented, the value given by Eq. (7) is approximate; nevertheless, size 9 is actually the maximum size found in the 15 topologies by using 50 observations (and size 12 is the maximum size found by using 200 observations) despite the presence of cancer progression trees of higher dimensions (Fig. 2c).

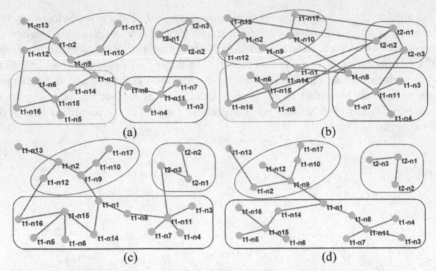

Fig. 6. The mapping of the identified RSs on the network reconstructed by zI3_graph, in the case of F2T1 topology, without noise. (a) 50 observations; (b) 100 observations; (c) 150 observations; (d) 200 observations. As the number of observations increases, it is possible to identify RSs of ever larger size. For example, it is possible to notice the expansion of the red RS in the passage from 50 to 100 observations, and the merging of two relatively large RSs in the passage from 100 to 150 observations. The RSs are always contained within the same cancer progression tree, and if they do not fully identify it (the green RS), they identify at least the main branches. The increase in the number of observations also gradually clarifies the underlying network, identified by the *zI_graph* method (Color figure online).

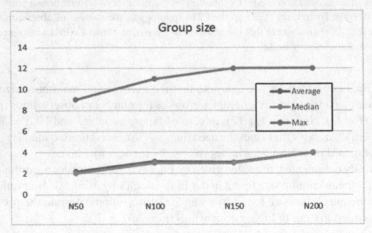

Fig. 7. The increase in the average, median and maximum size of the RSs identified in all topologies, as the number of observations increases *(zI_full*, with *zI_theta* = 3.0).

5 Conclusions

A common way of analyzing complex systems is that of thinking them as composed of numerous interacting parts in a non-linear way. Most frequently such relationships are thought of as binary, meaning they essentially involve two objects at a time. In this case, the result of the analysis consists in the reconstruction of the network of relationships between the parts of the system. Many useful algorithms and tools have been designed this way.

However, many systems for which this description is correct exhibit also emerging behaviors, and the formation of intermediate level structures between the mass of elementary components and the overall system is often observed. Intermediate-level structures, once formed, deeply affect the system, and therefore play a key role in understanding its behavior.

A consequent and interesting hypothesis is that the simultaneous use of both levels of description (the microlevel of the relationships between single entities, and the mesolevel constituted by their dynamically organized groups) can allow a better understanding of the phenomenon under examination. Therefore, in this work we are interested in taking advantage of the simultaneous use of different analyses of the same dataset. Within the family of RI tools we have the possibility to carry out analyses at both levels of description: in fact, the *zI-graph* method deals with the reconstruction of the network of binary relationships of the system under examination, whereas the result of the *zI_full* method consists in a list of dynamically coherent groups.

In order to verify the feasibility of the hypothesis we made use of simulated data, of which we know the ground truth; in particular the data refer to a situation concerning a significant situation, namely the dynamics of cancer progression. We applied probabilistic graphical models depicting the trends of accumulation of cancer driver events - organized in this work as progression trees - to produce cross-sectional mutational profiles of different patients. During the analysis we are interested in identifying the relationships between the various mutations and their overall organization.

The *zI_graph* method proved to be very effective. In this work we verified that the method, although not specifically designed for the analysis of cancer progression data, has performances comparable to those of one of the best state-of-the-art algorithms in this arena (CAPRI). Interestingly, the *zI_graph* method shows better performances in the presence of high noise levels. On the other hand, the *zI_full* method was able to identify the branches of cancer progression trees, and entire trees in case of sufficiently high amount of data.

Finally, we verified that the merger of the two levels of analysis leads to a much better reconstruction of the dynamic organization of the system under examination. The dynamic groups we have identified (the Relevant Sets) allow us to identify the branches of cancer progressions (if not the complete trees); the underlying network of relationships allows us to make hypotheses for the union of branches (in search of larger trees); the projection of the RSs on the network of relationships allows the identification of links that have lower probability of corresponding to real relationships. The reconstruction of the underlying dynamics is therefore finer than that possible using the individual levels of analysis separately, and it allows us to better distinguish between hypotheses of connection and evident connections among the system's constituents.

In future work we therefore plan to apply the method to non-tree underlying topologies, as well as to data concerning real patients, and to deepen the connections between the results of our two-level analysis and hypernetwork representations.

Funding. This research was funded by Università degli Studi di Modena e Reggio Emilia (FAR2019 project of the Department of Physics, Informatics and Mathematics).

References

1. Bar-Yam, Y., McKay, S.R., Christian, W.: Dynamics of complex systems (Studies in nonlinearity). Comput. Phys. **12**(4), 335–336 (1998)
2. Lane, D., Pumain, D., van der Leeuw, S.E., West, G.: Complexity Perspectives in Innovation and Social Change (Vol. 7). Springer Science & Business Media (2009). https://doi.org/10.1007/978-1-4020-9663-1
3. Lane, D.: Hierarchy, complexity, society. In: Hierarchy in Natural and Social Sciences, pp. 81–119. Springer, Dordrecht (2006). https://doi.org/10.1007/1-4020-4127-6_5
4. Ladyman, J., Lambert, J., Wiesner, K.: What is a complex system? Eur. J. Philos. Sci. **3**(1), 33–67 (2013)
5. Tononi, G., Sporns, O., Edelman, G.M.: A measure for brain complexity: relating functional segregation and integration in the nervous system. Proc. Natl. Acad. Sci. **91**(11), 5033–5037 (1994)
6. Tononi, G., McIntosh, A.R., Russell, D.P., Edelman, G.M.: Functional clustering: identifying strongly interactive brain regions in neuroimaging data. Neuroimage **7**(2), 133–149 (1998)
7. Hordijk, W., Steel, M.: Detecting autocatalytic, self-sustaining sets in chemical reaction systems. J. Theor. Biol. **227**(4), 451–461 (2004)
8. Villani, M., Filisetti, A., Benedettini, S., Roli, A., Lane, D., Serra, R.: The detection of intermediate-level emergent structures and patterns. In: ECAL 2013: The Twelfth European Conference on Artificial Life, pp. 372–378. MIT Press (2013)
9. Villani, M., Roli, A., Filisetti, A., Fiorucci, M., Poli, I., Serra, R.: The search for candidate relevant subsets of variables in complex systems. Artif. Life **21**(4), 412–431 (2015)
10. Bazzi, M., Porter, M.A., Williams, S., McDonald, M., Fenn, D.J., Howison, S.D.: Community detection in temporal multilayer networks, with an application to correlation networks. Multiscale Model. Simul. **14**(1), 1–41 (2016)
11. Holme, P., Saramäki, J.: Temporal networks as a modeling framework. In: Temporal Networks, pp. 1–14. Springer, Berlin, Heidelberg (2013). https://doi.org/10.1007/978-3-642-36461-7_1
12. D'Addese, G., Casari, M., Serra, R., Villani, M.: A fast and effective method to identify relevant sets of variables in complex systems. Mathematics **9**(9), 1022 (2021)
13. Margolin, A.A., et al.: ARACNE: an algorithm for the reconstruction of gene regulatory networks in a mammalian cellular context. In: BMC Bioinformatics, Vol. 7, No. 1, pp. 1–15. BioMed Central (2006)
14. Ramazzotti, D., et al.: CAPRI: efficient inference of cancer progression models from cross-sectional data. Bioinformatics **31**(18), 3016–3026 (2015)
15. Villani, M., et al.: A relevance index method to infer global properties of biological networks. In: Italian Workshop on Artificial Life and Evolutionary Computation, pp. 129–141. Springer, Cham (2017). https://doi.org/10.1007/978-3-319-78658-2_10
16. Villani, M., et al.: An iterative information-theoretic approach to the detection of structures in complex systems. Complexity 2018 (2018)
17. D'Addese, G., Sani, L., La Rocca, L., Serra, R., Villani, M.: Asymptotic information-theoretic detection of dynamical organization in complex systems. Entropy **23**(4), 398 (2021)

18. Righi, R., Roli, A., Russo, M., Serra, R., Villani, M.: New paths for the application of DCI in social sciences: theoretical issues regarding an empirical analysis. In: Italian Workshop on Artificial Life and Evolutionary Computation, pp. 42–52. Springer, Cham (2016). https://doi.org/10.1007/978-3-319-57711-1_4

19. Roli, A., Villani, M., Caprari, R., Serra, R.: Identifying critical states through the relevance index. Entropy 19(2), 73 (2017)

20. Sani, L., Lombardo, G., Pecori, R., Fornacciari, P., Mordonini, M., Cagnoni, S.: Social relevance index for studying communities in a facebook group of patients. In: International Conference on the Applications of Evolutionary Computation, pp. 125–140. Springer, Cham (2018). https://doi.org/10.1007/978-3-319-77538-8_10

21. Roli, A., Villani, M., Filisetti, A., Serra, R.: Dynamical criticality: overview and open questions. J. Syst. Sci. Complexity 31(3), 647–663 (2018)

22. Filisetti, A., Villani, M., Roli, A., Fiorucci, M., Serra, R.: Exploring the organisation of complex systems through the dynamical interactions among their relevant subsets. In: European Conference on Artificial Life 2015, pp. 286–293. MIT Press (2015)

23. Sani, L., et al.: Efficient search of relevant structures in complex systems. In: Conference of the Italian Association for Artificial Intelligence, pp. 35–48. Springer, Cham (2016). https://doi.org/10.1007/978-3-319-49130-1_4

24. Johnson, J.: Hypernetworks in the Science of Complex Systems, Vol. 3. World Scientific (2013)

25. Cover, T.M., Thomas, J.A.: Elements of Information Theory, 2nd edn. Wiley (2006)

26. Lachmann, A., Giorgi, F.M., Lopez, G., Califano, A.: ARACNe-AP: gene network reverse engineering through adaptive partitioning inference of mutual information. Bioinformatics 32(14), 2233–2235 (2016)

27. https://github.com/califano-lab/ARACNe-AP Accessed 17 July 2021

28. De Sano, L., et al.: TRONCO: an R package for the inference of cancer progression models from heterogeneous genomic data. Bioinformatics 32(12), 1911–1913 (2016)

29. Suppes, P.: A probabilistic theory of causality. British Journal for the Philosophy of Science 24(4), 409–410 (1973)

30. Caravagna, G., et al.: Algorithmic methods to infer the evolutionary trajectories in cancer progression. Proc. Natl. Acad. Sci. 113(28), E4025–E4034 (2016)

31. Angaroni, F., Chen, K., Damiani, C., Caravagna, G., Graudenzi, A., Ramazzotti, D.: PMCE: efficient inference of expressive models of cancer evolution with high prognostic power. arXiv preprint arXiv:1408.6032 (2014)

32. Merlo, L.M., Pepper, J.W., Reid, B.J., Maley, C.C.: Cancer as an evolutionary and ecological process. Nat. Rev. Cancer 6(12), 924–935 (2006)

33. Nowell, P.C.: The clonal evolution of tumor cell populations. Science 194(4260), 23–28 (1976)

34. Hanahan, D., Weinberg, R.A.: Hallmarks of cancer: the next generation. Cell 144(5), 646–674 (2011)

35. Fisher, R., Pusztai, L., Swanton, C.: Cancer heterogeneity: implications for targeted therapeutics. Br. J. Cancer 108(3), 479–485 (2013)

36. Meldrum, C., Doyle, M.A., Tothill, R.W.: Next-generation sequencing for cancer diagnostics: a practical perspective. The Clinical Biochemist Reviews 32(4), 177 (2011)

37. Beerenwinkel, N., Schwarz, R.F., Gerstung, M., Markowetz, F.: Cancer evolution: mathematical models and computational inference. Syst. Biol. 64(1), e1–e25 (2015)

38. Schwartz, R., Schäffer, A.A.: The evolution of tumour phylogenetics: principles and practice. Nat. Rev. Genet. 18(4), 213–229 (2017)

39. Sani, L., D'Addese, G., Graudenzi, A., Villani, M.: The detection of dynamical organization in cancer evolution models. In: Italian Workshop on Artificial Life and Evolutionary Computation, pp. 49–61. Springer, Cham (2019). https://doi.org/10.1007/978-3-030-450 16-8_6

40. Ma, X., et al.: Analysis of error profiles in deep next-generation sequencing data. Genome Biol. **20**(1), 1–15 (2019)

41. Powers, D.M.: Evaluation: from precision, recall and F-measure to ROC, informedness, markedness and correlation. arXiv preprint arXiv:2010.16061 (2020)

Artificial Chemical Neural Network for Drug Discovery Applications

Stefano Piotto[1,2(✉)] ⓘD, Lucia Sessa[1,2] ⓘD, Jacopo Santoro[1], and Luigi Di Biasi[1,3] ⓘD

[1] Department of Pharmacy, University of Salerno, Via Giovanni Paolo II, 132, 84084 Fisciano, SA, Italy
piotto@unisa.it
[2] Bionam Research Center for Biomaterials, University of Salerno, Via Giovanni Paolo II, 132, 84084 Fisciano, SA, Italy
[3] Department of Computer Sciences, University of Salerno, Via Giovanni Paolo II, 132, 84084 Fisciano, SA, Italy

Abstract. The drug design aims to generate chemical species that meet specific criteria, in-cluding efficacy against a pharmacological target, good safety profile, appropriate chemical and biological properties, sufficient novelty to ensure intellectual proper-ty rights for commercial success, etc. Using new algorithms to design and evalu-ate molecules in silicon de novo drug design is increasingly seen as an effective means of reducing the size of the chemical space to something more manageable for identifying chemogenomic research tool compounds and for use as starting points for hit-to-lead optimization.

Keywords: Convolutional neural network · Virtual screening

1 Introduction

"What gets us into trouble is not what we don't know. It's what we know for sure that just ain't so." Mark Twain.

The idea of taking experimental values and using a "descriptor set" for regression goes back to Hammett's pioneering formula linking reaction rates and equilibrium constants for reactions of benzene derivatives and computer-aided identification and quantification of the physicochemical properties of bioactive substances.

An increasing number of medicinal chemists have since applied various artificial intelligence methods [1, 2] to address the fundamental challenge of assessing and predicting the biological effects of chemicals. Neural networks were introduced in the 1990s due to their use as pattern recognition engines. In 1994, the first fully automat-ed molecular design method based on neural networks and evolutionary algorithms was published [3]. With time, various machine learning algorithms have been devel-oped and applied to drug design to help bridge the gap between chance-based and rational drug design [4, 5]. As with all such models, they suffer from the "garbage in, garbage out" problem, and they all struggle in some way with the most consistent challenge: which molecular features need to be put together for the most accurate predictions.

J. J. Schneider et al. (Eds.): WIVACE 2021, CCIS 1722, pp. 225–229, 2022.
https://doi.org/10.1007/978-3-031-23929-8_21

The ability of some deep learning methods to explore and predict complex relationships between molecular representations and observations provides a strong basis for optimism that these tools can produce more useful, generalizable insights.

Pharmaceutical companies that previously watched modern A.I. from the sidelines are now stepping in, with several large-scale collaborations between leading pharmaceutical and A.I. companies announced in recent years.

Virtual screening has been successfully used to optimize the hit finding process to reduce the cost of drug discovery while increasing its efficiency and predictability [6, 7] (Fig. 1).

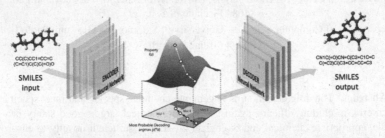

Fig. 1. An illustration of the autoencoder model for molecular design, including the joint property prediction model. Adapted from [8].

With strong generalization and learning capabilities, machine learning models that implement A.I. methods have been successfully applied in several aspects of the virtual screening process.

Structure-based virtual screening (SBVS) aims to search and rank the accessible chemical space for potential ligands based on a three-dimensional structural model of the given target macromolecule. Most molecular splicing studies have been performed with fully flexible ligands, and the objective is limited because of the computational cost of this method, although modern hardware has made it a possible competition on a large scale.

In recent years, A.I. algorithms have been introduced into SBVS by building non-parametric scoring functions. The review by Ballester et al. includes [9] the use of machine learning regression algorithms to develop AI-based scoring functions to improve protein-ligand binding affinity prediction.

Apart from designing and evaluating molecules in silico, de novo drug design is increasingly recognized as an effective tool for reducing the breadth of the chemical space to something more manageable for the identification of tool compounds for chemogenomic research and use as starting points for hit-lead optimization [10, 11]. In addition to designing and evaluating molecules in silico, de novo drug design is increasingly recognized as an effective tool to confront the space of possible molecules. The vastness of the chemical space (>1060 molecules) makes any attempt at a brute force approach impractical, and de novo drug design is the tool of choice for chemogenomic research and use as a starting point for hit-lead optimization.

Early de novo drug design methods used almost exclusively structure-based methods to develop ligands in binding pocket constraints for the target of interest, whether adapted directly from protein structure is often inferred from the properties of known ligands. One limitation of these early methods is that the generated structures tend to be synthetically unfeasible and have poor drug properties.

Another approach, called "inverse QSAR," treats the de novo design task differently. QSAR tries to find a clear inverse map Y→X from the attributes Y to the molecule descriptor space X and then backmaps from the favorable region in the X descriptor space to the corresponding molecules.

Many de novo drug design methods use molecular building blocks or fragments of synthetic compounds for molecular assembly to reduce the risk of creating unfavorable chemical structures. A chemical language model based on SMILES was created to solve the problem of chemically unfavorable structures.

2 Approach

The approaches described above are certainly useful, but they culpably forget some essential aspects of designing a drug. Mainly the definition of the mechanism of action is lacking. Recently, we have clarified the mechanism of action of a molecule, 2-hydroxy oleic acid, which has demonstrated high efficacy in treating gliomas and astrocytoma [12, 13]. This molecule is incorporated into lipid by the Ceramide synthase, and the resulting hydroxylated lipids are distributed to several finely regulated receptors.

The concentrations of all lipids in the cell membrane are finely regulated. This is essential for the cell to accurately control the membrane physical state (MPS), which is a function, not unique, of lipid concentrations. This implies that there can be infinite sets of concentrations that can achieve a given membrane physical macrostate. The MPS simultaneously affects dozens or hundreds of protein receptors and, for this reason, is a key aspect of drug design. According to many researchers [14, 15], a tumor state is a steady state different from a physiological state.

A physiological state can be defined as the set of concentrations of small molecules, receptors and their rate constants, and the MPS (Fig. 2).

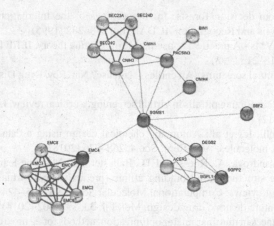

Fig. 2. Portion of the receptor network dedicated to fatty acid metabolism.

In the approach we have applied, we analyze the effect of binding a library of ligands against a network of receptors. For each ligand-receptor interaction, rate constants are estimated, and, consequently, the concentration trends of all species present are simulated. All interactions that lead to the collapse of a state, i.e., that reduce the concentrations of one or more biomarkers to 0, are discarded. All interactions that lead to the restoration of a physiological steady state are chosen as lead compounds. All molecules that, thanks to the interaction with multiple receptors, are able to bring the cell to a physiological steady state can be chosen as lead compound. It is important to emphasize that the interactions to which we refer are not limited to protein receptors but are, more generally, with any structure capable of changing state and transferring this information to other structures. With this definition of receptor, much broader than the traditional one, we can also consider cell membranes that, interacting with some molecules change their physical state [16, 17], aptamers, and the multifaceted galaxy of short RNAs.

Computational methods have become sophisticated enough to allow extensive screening of a library of molecules against an entire set of receptors. The network response is, of course, not linear. The number of simulations and calculations required is several orders of magnitude greater than what is faced today.

Albert Einstein said that "everything should be made as simple as possible, but no simpler," and drug discovery is undoubtedly one of the best examples that can be chosen to illustrate the concept. We can now make a conceptual leap in drug discovery by considering the multiple interactions of a drug in a cell.

References

1. Kopecký, J., Boček, K., Vlachová, D.: Chemical structure and biological activity on m-and p-disubstituted derivatives of benzene [20]. Nature **207**, 981 (1965)
2. Wessel, M.D., Jurs, P.C., Tolan, J.W., Muskal, S.M.: Prediction of human intestinal absorption of drug compounds from molecular structure. J. Chem. Inf. Comput. Sci. **38**, 726–735 (1998)
3. Schneider, G., Schuchhardt, J., Wrede, P.: Artificial neural networks and simulated molecular evolution are potential tools for sequence-oriented protein design. Bioinformatics **10**, 635–645 (1994)
4. Ho, T.K.: Random decision forests. In: Proceedings of the International Conference on Document Analysis and Recognition, ICDAR, pp. 278–282 (1995)
5. Schölkopf, B.: SVMs - A practical consequence of learning theory. IEEE Intelligent Systems Their Appl. **13**, 18–21 (1998)
6. Schneider, G.: Virtual screening: An endless staircase? Nat. Rev. Drug Discovery **9**, 273–276 (2010)
7. Scior, T., et al.: Recognizing pitfalls in virtual screening: a critical review. J. Chem. Inf. Model. **52**, 867–881 (2012)
8. Gómez-Bombarelli, R., et al.: Automatic chemical design using a data-driven continuous representation of molecules. ACS Cent. Sci. **4**, 268–276 (2018)
9. Ain, Q.U., Aleksandrova, A., Roessler, F.D., Ballester, P.J.: Machine-learning scoring functions to improve structure-based binding affinity prediction and virtual screening. Wiley Interdisciplinary Reviews: Computational Molecular Science **5**, 405–424 (2015)
10. Schneider, G.: Future de novo drug design. Mol. Inf. **33**, 397–402 (2014)
11. Segall, M.: Advances in multiparameter optimization methods for de novo drug design. Expert Opin. Drug Discov. **9**, 803–817 (2014)

12. Sessa, L., Nardiello, A.M., Santoro, J., Concilio, S., Piotto, S.: Hydroxylated fatty acids: the role of the sphingomyelin synthase and the origin of selectivity. Membranes **11**, 787 (2021)
13. Piotto, S., Sessa, L., Iannelli, P., Concilio, S.: Computational study on human sphingomyelin synthase 1 (hSMS1). Biochim. Biophys. Acta **1859**, 1517–1525 (2017)
14. Huang, S., Ernberg, I., Kauffman, S.: Cancer attractors: a systems view of tumors from a gene network dynamics and developmental perspective. Semin. Cell Dev. Biol. **20**, 869–876 (2009)
15. Serra, R., Villani, M., Barbieri, A., Kauffman, S.A., Colacci, A.: On the dynamics of random Boolean networks subject to noise: attractors, ergodic sets and cell types. J. Theor. Biol. **265**, 185–193 (2010)
16. Lopez, D.H., et al.: 2-Hydroxy arachidonic acid: a new non-steroidal anti-inflammatory drug. PLoS ONE **8**, e72052 (2013)
17. Piotto, S., et al.: Differential effect of 2-hydroxyoleic acid enantiomers on protein (sphingomyelin synthase) and lipid (membrane) targets. Biochimica et Biophysica Acta (BBA) - Biomembranes 1838, 1628–1637 (2014)

Trade and Finance

Information Flow Simulations in Multi-dimensional and Dynamic Systems

Riccardo Righi[1,2(✉)]

[1] European Commission, Joint Research Centre (JRC), Seville, Spain
`riccardo.righi@ec.europa.eu`
[2] Edificio EXPO, Calle Inca Garcilaso, 3, 41092 Seville, Spain

Abstract. The relevance of nodes with respect to the position they have in a network is often investigated with centrality measures. In particular, in cases where it is specifically meaningful to consider nodes' ability to cumulate and convey information, like in economic systems, betweenness centrality is one of the most pertinent options because of its underlying concept. However, this statistic presents two limitations. First, as it relies on the computation of shortest paths, it is grounded on a binary topological evaluation: every time a node is not located in the shortest path between two other nodes, it gains no score at all in its centrality (even if it is located on a path just one step longer). Second, betweenness centrality does not allow the direct analysis of multi-dimensional and dynamic networks: it has to be computed one dimension and one instant at a time, and this causes problems of comparability in case of weighted connections. The scope and the originality of this work is to design a network model that makes it possible to solve these issues. The proposed Dynamic Multi-Layer Network (DMLN) allows the structural representation of the multi-dimensional and dynamic properties of nodes' interactions. Then, this allows the computation of a metric that, based on Infomap random walks, assesses the level of information cumulated and conveyed by nodes in any moment and in any dimension of interaction. Importantly, this is performed without relying on a binary evaluation, and by jointly taking into account what occurred in all the dimensions and during the entire period, in which the system is observed. We present and discuss an implementation based on ICT worldwide trade of goods and services in the period 2004–2014.

Keywords: Multi-layer networks · Dynamic networks · Weighted networks · Infomap · Information flow simulations

1 Introduction

In the investigation of economic systems, especially if related to innovation dynamics [1,7–9], a crucial role is assumed by those agents that are able to

---

The views expressed are purely those of the author and may not in any circumstances be regarded as stating an official position of the European Commission.

© The Author(s) 2022
J. J. Schneider et al. (Eds.): WIVACE 2021, CCIS 1722, pp. 233–248, 2022.
https://doi.org/10.1007/978-3-031-23929-8_22

cumulate and convey relevant amount of information. As this is a matter of interactions, and therefore of the connective structure in which agents are embedded, network theory makes it possible to evaluate their topological position with centrality measures. From a theoretical point of view, among the several statistics that may be computed, betweenness centrality [6] appears to be one of the most pertinent to be computed for the investigation of the kind of agents' role that is here considered. Indeed, as it counts to how many shortest paths nodes belong, it allows the assessment of agents' position with respect to efficient transmissions of information occurring in the system. In this sense, betweenness centrality can be considered as a pertinent statistic to measure the level of the control that an agent may exert on the communication involving other nodes of the network: the larger is the number of shortest paths in which the agent is located, the more it can rule the flows of information, so eventually forcing other nodes to communicate via less efficient, i.e., longer, paths. Also, the more information passes through an agent, the more its knowledge is likely to increase.

Despite its usefulness, betweenness centrality—which can also be computed for weighted networks [4]—presents two limitations. First, it is based on the computation of shortest paths, and this means to assess in a binary way the inclusion (or not) of nodes in the set of nodes involved in the sequence of connections minimizing the number of steps that are needed to move between couples of nodes. It follows that every time a node is not exactly included in the shortest path between two other nodes, it gains no score at all in its centrality (even if it is located on a path which is just one step longer). For instance, if we consider two nodes i_1 and i_2, and we want to check if nodes i_3, i_4 and i_5 are central with respect to them (in a shortest paths perspective), we can face the situation described in Fig. 1: i_3 is located in the shortest path, and thus it is considered as central, while i_4 and i_5 are not located in the shortest path, and so are considered as not central at all. In addition, what is interesting to observe is that the shortest path that connects i_1 and i_2 passing through i_4 is shorter than the one that connects i_1 and i_2 passing through i_5 (the former has a length equal to 3, while the latter a length equal to 4). Although it can be argued that i_4 is more central than i_5 with respect to the couple of nodes i_1 and i_2, in terms of shortest path there is no difference at all. For this reason, the computation of a metric that does not rely on such a dichotomic approach would allow a finer consideration and measurement of nodes' ability to cumulate and convey information given the topology of the network.

Second, betweenness centrality does not allow the direct analysis of multidimensional and dynamic networks [2, 3, 10]. Indeed, in case of multiple types of interactions and of multiple instants over time, it has to be separately computed on the subnetworks determined by considering one type of interaction and one instant at a time. In case of a weighted network structure, this creates problems of comparability, as it can be pertinently argued that the role that an agent/node has with respect to a specific type of interaction and/or in a specific moment in time, also depends on the weight that the corresponding subnetwork has in the overall system. For instance, if we consider a network of individual interactions

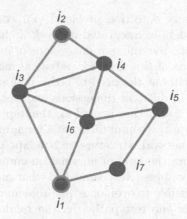

Fig. 1. Simple network with 7 nodes. The shortest path between nodes i_1 and i_2, which is colored in orange as well as the two aforementioned nodes, passes through i_3 and has a length of 2 steps. The shortest path connecting i_1 and i_2 and necessarily passing through i_4, has a length of 3 steps (either $i_1 \rightarrow i_6 \rightarrow i_4 \rightarrow i_2$, or $i_1 \rightarrow i_3 \rightarrow i_4 \rightarrow i_2$). The shortest path connecting i_1 and i_2 and necessarily passing through i_5, has a length of 4 steps (either $i_1 \rightarrow i_6 \rightarrow i_5 \rightarrow i_4 \rightarrow i_2$, or $i_1 \rightarrow i_7 \rightarrow i_5 \rightarrow i_4 \rightarrow i_2$). Even if the lengths are different, which can lead to the conclusion that i_4 is closer than i_5 to the couple of nodes i_1 and i_2, there is no difference between them: as they are simply not located in the shortest path connecting i_1 and i_2, for both of them the betweenness centrality gains no score at all. (Color figure online)

in different personal contexts, it is certainly different to be on the shortest path connecting two friends that are usually met a couple of times per year, from being in the shortest path connecting two colleagues that are responsible of two separate tasks of a same project. In the latter cases, it is very likely that the amount of information to be managed is much larger, and this can be used as an argument to state that in the second case the agent has a more relevant role. Similarly, but with respect to time, if we consider a network made of airports and flight connections, the relevance of an airport certainly changes from being in the shortest path connecting two cities during a normal period of the year, or being in the same situation but during winter holidays, when the volume of people moving from one place to the other sensibly increases. Also in this case, the amount of information that the considered airport has to manage in the latter situation is indeed more considerable, and again this point can be used as an argument to state the in the second circumstances the airport has a more important role.

In this work we design a network structure allowing the computation of a statistic that can assess the amount of information passing through each node, and that makes it possible to overcome the previously described limitations of betwenness centrality. More specifically, being inspired by the analysis of multi-layer networks [5] and by the Infomap algorithm [11] (which was initially developed for a community detection problem), our contribution is about the definition of a single and fully integrated network architecture that is able to structurally represent both the different dimensions of interactions, and also the passing of time. By means of this network statistical model, the system can be considered as a whole, without the need to break it in subnetworks in order to assess the role of agents in different dimensions and instants of time. Once it is set, the proposed network structure is used for the implementation of Infomap. As this algorithm relies on the computation of a large number of random walks moving throughout the network structure, it is adapt to simulate information flows, and thus to measure the level of information cumulated and conveyed by nodes. Although this may appear very similar to what measured by betweenness centrality, it makes it possible to overcome the problem of the binary evaluation of agents with respect of shortest paths. As the random walks move according to probabilistic principles, no path between nodes is a priori excluded and, therefore, the relevance of nodes with respect to the topology of the network is evaluated in a way which is more gradual than the one considering the location (or not) in shortest paths.

The work is structured as follows. In the Sect. 2 we define our approach, and in Sect. 3 we discuss its implementation for the analysis of the ICT worldwide trade of goods and services in the period 2004–2014, a dataset describing yearly observations about 5 categories of goods and 2 of services.

2   Network Statistical Model

The initial multi-dimensional network G, i.e., a network whose links are of different types, can be defined as

$$G = (D, V, E) \tag{1}$$

where

- $D = \{1, \ldots, \alpha, \ldots, \Omega\}$ is the set of dimensions, and α represent the α-th dimension, Ω is the number of dimensions. Hence $|D| = \Omega$.
- $V = \{1, \ldots i, \ldots, I\}$ is the set of nodes, and i represent the i-th node, so that I is the number of nodes. Hence $|V| = I$. As each node of the network can take part in multiple dimensions, according to literature [5] we distinguish between (i) the notion of *physical-node*, which is used to indicate what in network analysis is usually indicated with the term *node* (i.e., an agent of the system), and (ii) the notion of *state-node* to refer to the *physical-node*'s projection on a specific dimension. Therefore, we will refer to the *physical-node* $i \in V$ active in dimension $\alpha \in D$, as the *state-node* i_α.
- $E = \{e_{i_\alpha, j_\beta} : G_{i_\alpha, j_\beta} = 1\}$ is the set of links e_{i_α, j_β}, active from the *state-node* i_α to the *state-node* j_β. Alternatively, we refer to the edge e_{i_α, j_β} as $\overline{i_\alpha j_\beta}$. It is important to highlight that the connections between *state-nodes* are used to allow the representation of interactions in multiple dimensions.

The graph G is weighted, and the weight of the edge $\overline{i_\alpha j_\beta}$ is noted as $\omega_{i_\alpha, j_\beta}$.

2.1   Information Drift Across Multiple Dimension

Considering the elements previously described, any edge $\overline{i_\alpha j_\beta}$ can exist in the following cases:

i) $i \neq j \wedge \alpha = \beta$, which hence defines the intra-dimension connection $\overline{i_\alpha j_\alpha}$ involving the two different *physical-nodes* i and j located in a same dimension,

ii) $i = j \wedge \alpha \neq \beta$, which hence defines the inter-dimension connection $\overline{i_\alpha i_\beta}$ involving a single *physical-node* $(i = j)$, but referring to two distinct *state-nodes* of it,

iii) $i = j \wedge \alpha = \beta$, which hence defines the *state-node* self-loop $\overline{i_\alpha i_\alpha}$, since it involves a single *physical-node* $(i = j)$ in a single dimension $(\alpha = \beta)$.

Regarding the missing combination, i.e., the one with $i \neq j \wedge \alpha \neq \beta$, we impose that $e_{i_\alpha, j_\beta} = 0 \; \forall \; i \neq j \wedge \alpha \neq \beta$. Therefore, no inter-dimensional connection involving *state-nodes* of distinct *physical-nodes* is allowed. Even if it is statistically possible to define such edges, no reason to allow the existence of this kind of connections is identified by the authors. As the current model is aimed at representing the spread of information, we can say that it is always needed a propagation medium (or a channel) across which information can move from an initial point to an end point. The propagation medium can be either a *physical-node* $i \in V$ (in the sense two *state-nodes* can interact if they are referred to a same *physical-node*), or a dimension $\alpha \in D$ (in the sense that two *physical-nodes* can interact when they have *state-nodes* located in a same dimension). As information circulates by means of interactions, the movement of information from agent i to agent j can only take place if the *physical-nodes* are both active on a common dimension $\alpha \in D$. And the movement of information from dimension α to dimension β can only be carried out by a *physical-node* $i \in V$ that has *state-nodes* active in α and β. The case of self-loops, does not present any incompatibility with what just stated.[1] We label the connections listed above as (i) *information exchanges*, (ii) *information dimensional-switches*, and (iii) *information holding*, respectively.

2.2   Information Drift Across Time

We now assume that G is dynamic, i.e., it is observable in a sequence of instances $t \in T = \{0, 1, t, ..., T_{max}\}$, where $T \subset \mathbb{N}$ and T_{max} is the last instant. As we consider D like a set of dimensions characterizing the network G, we consider T as a set of instants in which the network G can be observed. Time crosses all the existing Ω dimensions of the network G and generates a series of projections of the same dimensions. In other words, we need to indicate the instant in time in which any dimension $\alpha \in D$ is observed. This implies that the *state-nodes* no

---

[1] Self-loops can be seen as self-interactions, or as an information holding. Despite not involving different *state-nodes*, this type of connection influences the distribution of the information gathered by the nodes.

longer depend exclusively on the dimension in which *physical-nodes* are considered, but also on the instant t of time in which the *physical-nodes* are observed. Hence, the *state-node* i_α becomes $i_{\alpha,t}$. In addition, the definition of the edges has to be accordingly adapted. In order to locate connection over time, we define them as $e_{i_{\alpha,\hat{t}},j_{\beta,\tilde{t}}}$, or alternatively $\overline{i_{\alpha,\hat{t}}\,j_{\beta,\tilde{t}}}$, where $\hat{t},\tilde{t} \in T$.

In case that $\hat{t} = \tilde{t}$, which means that the considered link connects two *state-nodes* located in the same instant in time, all the considerations regarding intra- and inter- connections and self-loops remain unvaried with respect to what has been discussed in Sect. 2.1. On the other hand, in case $\hat{t} \neq \tilde{t}$, some considerations have to be added as follows.

Directionality Through Time. First, time is a peculiar dimension in the sense that the movement through time is allowed in only one direction, i.e., from past to future. Therefore, when considering inter-temporal *state-nodes* connections we necessarily have to impose only one direction to the flow of information. Formally, $\forall\ \hat{t}. < \tilde{t}$ the only possible connection between $i_{\alpha,\hat{t}}$ and $j_{\beta,\hat{t}}$ is $\overrightarrow{i_{\alpha,\hat{t}}\,j_{\beta,\tilde{t}}}$, with the arrow pointing to the same direction of time.

Flow over Time: Straight and Bending Inter-temporal Movement. Second, similarly to what discussed in Sect. 2.1 about the connections between two *state-nodes* referred to two distinct dimensions, also the connection between *state-nodes* referred to two distinct dimensions and located in distinct moments in time is not allowed. We therefore allow inter-temporal connections only when the same *physical-node* (but in different moment in time) is involved. Following this condition, and also always under the condition that $\hat{t} < \tilde{t}$, two possibilities remain:

a) the first is that $\alpha = \beta$. The connection can so be re-written as $\overrightarrow{i_{\alpha,\hat{t}}\,i_{\alpha,\tilde{t}}}$ and it identifies a connection between two different time projections of a same *physical-node* in a same dimension. As information remains attached to the same *physical-node* i in the same dimension α and it just moves over time, we refer to this type of connections as *information straight inter-temporal movements*,

b) the second is that $\alpha \neq \beta$. This connection, that can be written as $\overrightarrow{i_{\alpha,\hat{t}}\,i_{\beta,\tilde{t}}}$, is between two *state-nodes* of a same *physical-node* i, but it is referred to different dimensions, α and β, in two distinct moments in time, $\hat{t}$ and $\tilde{t}$. This connection represents information moving over time and dimensions by means of the same *physical-node*. We refer to this type of connections as *information bending inter-temporal movements*.

Information Time Lag. Third, a crucial point regarding movement over time has to be discussed. Information can theoretically move from one instant to any following instants. As long as the involved agent keeps memory of some information acquired in time $\hat{t}$, it can use it with any lag $\psi \in \mathbb{N}^+$, i.e., it can use it in any future instant. In other words, when defining inter-temporal connections, the lag between $\hat{t}$ and $\tilde{t}$ can be greater or equal than 1. Formally, $\psi(\hat{t},\tilde{t}) \geq 1$.

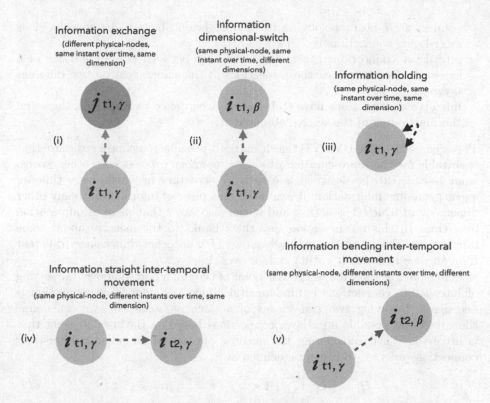

Fig. 2. Schematic representation of the type of information movements considered in the proposed statistical model and based on the definition of *state-nodes*. Letters i and j indicate two distinct agents of the system (i.e., two distinct *physical-nodes*), greek letters α and β indicate two different dimensions of interaction, and $t1$ and $t2$ indicate two instants over time (with $t1$ occurring before $t2$).

2.3   The Dynamic Multi-layer Network (DMLN) Structure

In order to statistically represent a multi-dimensional and dynamic system crossed by information flows, we define in this work a new dynamic multi-layer network (DMLN) structure. The core elements of the DMLN structure are:

- in order to represent the time projections of the dimensions included in the network G, we use network layers. More specifically, any combination of $\alpha \in D$ and $t \in T$ identifies a distinct layer of the network,
- with respect to the types of connections discussed in Sect. 2.2 and represented in Fig. 2, namely (i) *information exchanges*, (ii) *information dimensional-switches*, (iii) *information holding*, (iv) *information straight inter-temporal movements*, and (v) *information bending inter-temporal movements*, the first one and the third one of them have to be considered as intra-layers connections, as they take place on a same layer. The remaining ones, since they

connect *state-nodes* belonging to distinct layers, have to be considered as inter-layers connections,

- inter-layer connections can occur exclusively between two *state-nodes* of a same *physical-node* (i.e., the projections of the same agent on two different layers),
- inter-layer connections have to be set in accordance to some hypotheses on the functioning of the system observed.

The originality of the DMLN is that it makes it possible to define a structure that is suitable for the representation of agents' memory processes. In other words, what intended to be designed is a network structure in which every time an agent gets some information, it can move that piece of information to any other dimension in which it is active, and it can also carry that piece of information over time. In this sense, we can say that thanks to the memory about some information, the agent can take advantage of it in other dimensions (different from the one in which it got it) and/or over time.

Therefore, as reported in the last point of the list reported above, the setting of inter-layers connections is fundamental. Since any inter-layer connection is necessarily involving two *state-nodes* of a same *physical-node*, we can define the subsets all possible inter-layer connections based on the specific agent that is involved. When considering the agent i, the set of all possible inter-layer connections referred to it can be defined as

$$\Theta_i = \{ \overrightarrow{i_{\alpha,\hat{t}} \, i_{\beta,\tilde{t}}} \mid \hat{t} < \tilde{t} \lor (\hat{t} = \tilde{t} \land \alpha \neq \beta) \} \tag{2}$$

Any of these subsets describes all the possible paths through which information can spring from one layer[2] to another, by means of i-th agent of the system. In Fig. 3, a schematic representation of Θ_i for a network with 3 dimensions of interactions and 3 instants over time is shown. Clearly, depending on the system considered for the analysis, not necessarily all the possible inter-layer connections of each Θ_i have to be considered as active or pertinent. Therefore we define $\theta_i \subseteq \Theta_i$, in order to indicate which are the inter-layers connections that have to be finally considered for the specific analysis. It is important to recall that the complementary connections, i.e., the intra-layers connections, don't need any kind of discussion as the work developed here does not present any originality with previous works on multi-layer networks [5].

3   DMLN Analysis of ICT Worldwide Trade 2004–2014

The DMLN structure is implemented for the analysis of the ICT worldwide trade 2004–2014. The dataset consists of import and export flows (of goods and services) between worldwide countries, with the values of both imports and exports converted into constant prices euros using exchange rates by Eurostat.

---

[2] We recall that each layer is a determined by a specific combination of an instant in time and a type of interactive dimension.

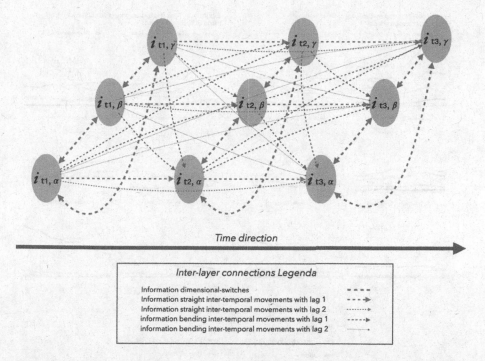

Fig. 3. Schematic representation of Θ_i, i.e., the set of all possible inter-layer connections related to agent i, in case of a DMLN with three dimensions (α, β, γ) and three instants over time $(t1, t2, t3)$. Each grey circle represents a *state-node* of the *physical-node i* (i.e., the projections of agent i in the different layers defined according to combinations of dimensions and instants in time). As Θ_i is exclusively about what potentially occurring in terms of inter-layer connections, the connections representing *information exchanges* and *information holding* are not considered. As they occur within the boundary of a same layer, they are intra-layer connections.

No threshold is used to sample the most relevant trade flows, hence all trade flows are considered. From an initial set of 45 geographical areas[3], in order to perform an exploratory analysis intended to investigate world patterns, the countries belonging to the European Union are considered as a single aggregate of 28 countries, i.e., the EU28[4]. By doing so, the final network considered is made of 18 distinct geographical areas. The time period that is considered goes from 2004

---

[3] Austria, Belgium, Bulgaria, Croatia, Cyprus, Czech Republic, Denmark, Estonia, Finland, France, Germany, Greece, Hungary, Ireland, Italy, Latvia, Lithuania, Luxembourg, Malta, Netherlands, Poland, Portugal, Romania, Slovak Republic, Slovenia, Spain, Sweden, United Kingdom, Norway, Switzerland, Australia, Brazil, Canada, China, India, Japan, Korea, Russia, Taiwan, United States, Africa, rest of American countries, rest of Asian countries, rest of European countries and rest of Oceanian countries.

[4] The analysis considers a period entirely before Brexit, hence we consider UK as still part of the European Union.

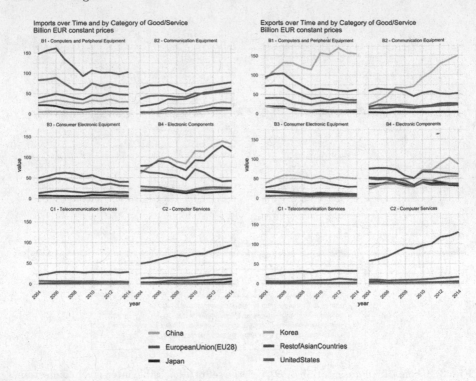

Fig. 4. Observed import and exports of ICT goods and services by category (category 'B5 - miscellaneous' is not reported) for main geographical areas.

to 2014, with yearly observation. Therefore, the system can be observed in 11 instants over time. Concerning the disaggregation by type of goods, the OECD Guide to measuring the Information Society (OECD, 2011) is followed. The ICT goods are defined at 6-digit level using the Harmonised System classification, aggregated into 70 blocks of items and organised in five product categories: (i) B1 - computers and peripheral equipment, (ii) B2 - communication equipment, (iii) B3 - consumer electronic equipment, (iv) B4 - electronic components, and (v) B5 - miscellaneous. For ICT services, the data has been organized according to the extended balance-of-payments categories (EBOPS) for each country. The ICT services categories used for the analysis are: (i) C1 - telecommunications services, and (ii) C2 - computer services. Therefore, the system has 7 dimensions of interaction. Import and export values for the six major geographical areas, i.e., China, EU28, Japan, (South) Korea, Rest of Asias Countries and US, are represented in Fig. 4.

Based on this data, the DMLN is built with the following characteristics:

– directionality of the connections going in reverse-way with respect to the flow of goods/services. For instance, China's export to Korea is represented as a connection that goes from Korea to China,

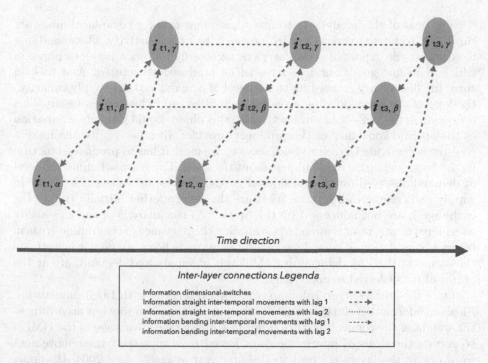

Fig. 5. Schematic representation of θ_i as modeled for the study of ICT trade flows 2004–2014. Only three dimensions (α, β, γ) and three instants over time $(t1, t2, t3)$ are represented for simplicity. Each grey circle represents a *state-node* of the *physical-node* i (i.e., the projections of geographical are i in the different layers defined according to combinations of goods/services categories and time).

- no self-loops (i.e., no *information holding*), because of data unavailability on products and services sold internally in each area,
- the *state-nodes* belonging to a same area and associated to different dimensions in a same year, are all interconnected in a bi-directional way (i.e., all possible *information dimensional-switches* are considered), as represented in Fig. 5, where the green lines are the only ones presenting a double arrowhead,
- 7 dimensions of interactions and 11 instants over time, for a total of 77 layers,
- no *information bending inter-temporal movements* and information time lag for *information straight inter-temporal movements* always equal to 1, as represented in Fig. 5,
- weight of intra-layer connections based on the value of imports and exports,
- all inter-layer connections with constant and very low weight. The fact that inter-layer connections' weight are constant makes it possible to maintain balance among all the layers. In addition, to set with very low values forces the algorithm to consistently explore the internal network structure of any layer (i.e., the intra-layer connections).

The goal of this analysis is to investigate how much geographical areas are supposed to cumulate know-how by means of their trade activity. The underlying idea is that the export of goods/services necessarily follows a previous phase in which the same goods/services have to be produced/structured. And this, in turn, implies to vary (according to the level of demand and the supply capacity) the level of the corresponding economic activities and industrial processes. This is crucial, in the sense that once activated, this phase should generate a variation of the specific know-how of the producer/provider. In other words, the more a country is demanded to sell a good/service, the more it has to produce/structure it, the larger expertise it cumulates about the same. The reasons behind the level of demand observed for the considered goods/services, as well as the ability to supply and commercialize them (that are the elements that initially trigger the exchanges), are not addressed by this work.[5] As the analysis of the case-study considered in this work is aimed at estimating the relevance of geographic areas in terms of cumulation of know-how generated by trade flows, we do not investigate the elements that, by determining the levels of supply and demand, are at the origin of the observed imports/exports.

Once the DLMN is set, Infomap algorithm [11] is run with 1,000 simulations. Flows of information run through the structure in a random way but according to the weight of the connections[6]. In order to entirely take advantage of the DMLN structure, the flows of information are forced to always start their movement from one of the layers referred to the first year available, i.e., 2004. By doing so, any simulated stream runs across all the periods considered in the DMLN and, when advancing to any new instant in time, the information flow can either move to any other layer of the same instant in time, or it can move to some layer referred to the next instant in time. Therefore, it can never go back to a previous instant in time.

The output of this analysis is the flow (in percentage terms) that any *state-node* has been crossed by. Therefore, the use of Infomap algorithm over the DMLN structure allows us to assess the percentage of information that any geographical area has cumulated in any instant over time and for any category of ICT goods/services. In order to discuss these preliminary results, we compare them with the computation of the weighted betweenness centrality (WBC) [4,6]. As previously discussed in Sect. 1, this statistic considers the number of times that a node falls in the shortest paths connecting couples of other nodes. It is therefore an indicator of the control that nodes can exert on the rest of the network. In terms of WBC, to be a central node means to be located in a relatively large number (depending in the system considered) of the most efficient connec-

---

[5] Typically, these reasons are related to differences in local production costs, or to a different availability of raw materials in some areas, or also to certain industrial specializations that have emerged over time.

[6] The larger the weight of a connection, the higher the probability that the flow goes through that connection.

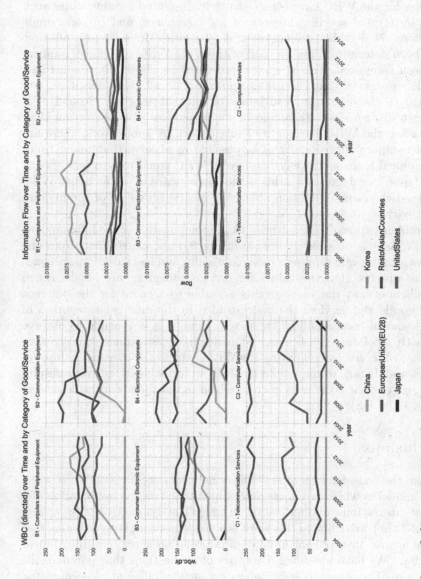

Fig. 6. Comparison of the computation of weighted betweenness centrality in the subnetworks determined by the combinations of categories of goods/services and year, vs. the computation of the Infomap flow based on the DMLN structure set for the analysis of ICT international trade in the period 2004–2014. Only six major geographical areas are considered in the plots and the values for the category 'B5 - miscellaneous' are not represented.

tions between other nodes[7]. Therefore, a central node in terms of WBC can rule the flows of information, so eventually forcing other nodes to communicate via less efficient, i.e., longer, paths. Also, a central node in WBC is facilitated in the to cumulation of knowledge, as much information is likely to pass through it.

The values for the WBC have been separately computed for any subnetwork determined by (i) one specific category of goods/services, and (ii) one single year. Therefore, 77 detached subnetworks (7 categories times 11 instants over time) have been determined, and in each of them the WBC of the geographical areas has been computed. This is what represented in Fig. 6, left panel. First, it is possible to observe that the values appear to be relatively sensitive, and from one year to the other they can jump (or fall) substantially. Second, it is not possible to discern if some category of goods/services is more relevant than the others: since the WBC is computed subnetwork by subnetwork, there are problems of comparability. Finally, it is possible to observe that many times some geographical areas are revealed to have a WBC equal to zero (e.g., South Korea and Japan in category B4): this is the consequence of the fact that WBC is based on a dichotomic evaluation about the location (or not) of the node in the shortest paths.

These problems appear to be solved with the computation of the Infomap flow in the DMLN, as it is possible to observe in Fig. 6, right panel. First, from one year to the other, the values present more continuity and less variations. Second, it possible to observe that the values for certain categories are larger, which therefore indicates that the whole results are able to account for the different amounts of goods and services that were traded in the different categories: in this sense, issues of comparability are solved. Finally, it is possible to observe that even with secondary roles, many countries have less values equal to zero (than for what observed by computing the WBC): this is the case for example of South Korea and Japan, whose role in the ICT sector is indeed relevant despite the smaller size of their economies in comparison to the three worldwide powers (China, EU28 and US).

4   Conclusions

Starting from the consideration of multi-layer networks, we have built a statistical network model in which layers are determined based on the combination of a dimension of interactions, and an instant in time. The dynamic multi-layer network (DMLN) that we define in this work is a single network structure aimed at representing agents' involvement in flows of information across multiple dimensions and time. We have identified five types of connections that populate the proposed model. These are: (i) *information exchanges* (different *physical-nodes*, same instant over time, same dimension); (ii) *information dimensional-switches*

---

[7] Betweenness centrality relies on the computation of the shortest paths, and any shortest path has to be intended as the most efficient connection (because it minimizes its length) between the considered couple of nodes.

(same *physical-node*, same instant over time, different dimensions); (iii) *information holding* (same *physical-node*, same instant over time, same dimension); (iv) *information straight inter-temporal movement* (same *physical-node*, different instants over time, same dimension); and (v) *information bending inter-temporal movement* (same *physical-node*, different instants over time, different dimensions).

An exploratory analysis of the network of international ICT trade from 2004–2014 is developed to test the implementation of Infomap algorithm on the DMLN structure. In order to evaluate the results, weighted betweenness centrality (WBC) is considered as a term of comparison. Indeed its underlying principles fit the investigation of economic systems, as WBC is adequate to assess the level of control that nodes exert on the exchanges/communications among other nodes. However, this statistic is affected by some issues: it based on a binary assessment about the location (or not) of nodes in the shortest paths, and it has problems of comparability when some subnetworks of the same weighted network are considered. The Infomap flow computed for the DMLN reveals values that are less sensitive and discontinuous on a time basis, that are more able to consider the role of nodes of secondary relevance, and that better reflect the original proportions between the weight of the different dimensions considered. In this sense, the proposed approach is able to solve the discussed limitations of one of the most used centrality measures, i.e., WBC.

Further developments of the work will include more accurate measurements and quantification of the aspects described above, and additional investigations on other dynamic and multi-dimensional networks. Moreover, as the Infomap algorithm was initially developed for a community detection problem, the investigation of communities emerging from the DMLN is also one of the next steps related to this work.

References

1. Arthur, W.B.: Foundations of complexity economics. Nat. Rev. Phys. **3**(2), 136–145 (2021)
2. Barrat, A., Barthelemy, M., Pastor-Satorras, R., Vespignani, A.: The architecture of complex weighted networks. Proc. Natl. Acad. Sci. **101**(11), 3747–3752 (2004)
3. Boccaletti, S., Latora, V., Moreno, Y., Chavez, M., Hwang, D.U.: Complex networks: structure and dynamics. Phys. Rep. **424**(4–5), 175–308 (2006)
4. Brandes, U.: A faster algorithm for betweenness centrality. J. Math. Sociol. **25**(2), 163–177 (2001)
5. De Domenico, M., Lancichinetti, A., Arenas, A., Rosvall, M.: Identifying modular flows on multilayer networks reveals highly overlapping organization in interconnected systems. Phys. Rev. X **5**(1), 011027 (2015)
6. Freeman, L.C.: Centrality in social networks conceptual clarification. Soc. Netw. **1**(3), 215–239 (1978)
7. Lane, D.: Complexity and innovation dynamics. Handb. Econ. Complex. Technol. Change **63** (2011)
8. Lane, D., Maxfield, R.: Building a new market system: effective action, redirection and generative relationships. Complex. Perspect. Innov. Soc. Change, 263–288 (2009)

9. Lane, D.A.: Innovation cascades: artefacts, organization and attributions. Philos. Trans. Roy. Soc. B **371**(1690), 20150194 (2016)
10. Newman, M.E.: The structure and function of complex networks. SIAM Rev. **45**(2), 167–256 (2003)
11. Rosvall, M., Bergstrom, C.T.: Maps of random walks on complex networks reveal community structure. Proc. Natl. Acad. Sci. **105**(4), 1118–1123 (2008)

Investigation
of the Ramsey-Pierce-Bowman Model

Johannes Josef Schneider[1]([✉])(iD), Andreas Lang[2], Christian Hirtreiter[3],
Ingo Morgenstern[3], and Rudolf Marcel Füchslin[1,4](iD)

[1] Institute of Applied Mathematics and Physics, School of Engineering, Zurich
University of Applied Sciences, Technikumstr. 9, 8401 Winterthur, Switzerland
`johannesjosefschneider@googlemail.com, scnj@zhaw.ch`
[2] ATS Systems Oregon, 2121 NE Jack London St., Corvallis, OR 97330, USA
[3] Physics Faculty, University of Regensburg, Universitätsstr. 31, 93053 Regensburg,
Germany
[4] European Centre for Living Technology, Ca' Bottacin, Dorsoduro 3911,
Calle Crosera, 30123 Venice, Italy
`https://www.zhaw.ch/en/about-us/person/scnj/`

Abstract. In this publication, the model by Ramsey, Pierce, and Bowman for finding the hierarchical order of the various sectors of an economy, conceiving each of them as users or suppliers for other sectors of the economy, is investigated. Computational results for a benchmark instance are provided.

Keywords: Simulated annealing · Threshold accepting · Ramsey ·
Spin glass · Traveling salesman problem

1 Introduction

In 1969, Ramsey, Pierce, and Bowman considered the problem to find a hierarchical order of the various sectors of an economy, conceiving each of them as users or suppliers of goods and services for other sectors, including itself [1]. For this purpose, an exchange matrix J between the various sectors is considered with $J(i, j) \geq 0$ being the value of all products of sector i which are used for manufacturing products of a specified normalized value in sector j. Examples of these sectors are banks and insurance companies, the automobile industry, the chemical industry, publishing companies, agriculture and fishing, transport, restaurants, but also non-profit organizations like churches, and the entertainment industry. The exchange matrix is generally asymmetric for data from real-world economies.

The problem the Ramsey-Pierce-Bowman (RPB) model treats is to find a hierarchical order σ of N economic sectors in the way that $\sigma(1)$ is the sector buying products of the largest value from other sectors, whereas $\sigma(N)$ is the sector providing the largest-valued supply of products to other sectors. Thus, $\sigma(1)$ is called the largest user, $\sigma(N)$ the largest supplier. To solve this problem,

J. J. Schneider et al. (Eds.): WIVACE 2021, CCIS 1722, pp. 249–259, 2022.
https://doi.org/10.1007/978-3-031-23929-8_23

a cost function or Hamiltonian to be minimized is defined in [1]:

$$\mathcal{H}_J(\sigma) = \sum_{i=1}^{N-1} \sum_{j=i+1}^{N} J(\sigma(i), \sigma(j)) = \sum_{i<j} J(\sigma(i), \sigma(j)). \tag{1}$$

This Hamiltonian exhibits some very interesting properties [1]:

- If J is a symmetric matrix, then $\mathcal{H}_J \equiv$ const, as for each permutation σ, either $\sigma(i) < \sigma(j)$ or $\sigma(i) > \sigma(j)$ for $i \neq j$. Thus, either $J(\sigma(i), \sigma(j))$ or $J(\sigma(j), \sigma(i))$ is part of the upper triangular matrix and adds to the total costs $\mathcal{H}_J(\sigma)$ of the configuration σ. If now $J(\sigma(i), \sigma(j)) = J(\sigma(j), \sigma(i))$ for a pair (i, j), it does not make any difference which of these two values is added to $\mathcal{H}_J(\sigma)$.
- Furthermore, the Hamiltonian $\mathcal{H}_J$ is additive with respect to the underlying exchange matrix J: let $J = K + L$, then

$$
\begin{aligned}
\mathcal{H}_{K+L}(\sigma) &= \mathcal{H}_J(\sigma) \\
&= \sum_{i<j} J(\sigma(i), \sigma(j)) \\
&= \sum_{i<j} (K(\sigma(i), \sigma(j)) + L(\sigma(i), \sigma(j))) \\
&= \sum_{i<j} K(\sigma(i), \sigma(j)) + \sum_{i<j} L(\sigma(i), \sigma(j)) \\
&= \mathcal{H}_K(\sigma) + \mathcal{H}_L(\sigma).
\end{aligned} \tag{2}
$$

Combining these two properties, the aim of the optimization process can be described more precisely: let

$$K(i, j) = \min\{J(i, j), J(j, i)\} \tag{3}$$

and

$$L(i, j) = J(i, j) - K(i, j). \tag{4}$$

As K is symmetric, $\mathcal{H}_K \equiv$ const is a flow amount independent of the hierarchical order of the economic sectors in the exchange matrix and thus plays no role in the optimization process. Please note that the original optimization problem to minimize $\mathcal{H}_J$ is thus equivalent to the minimization of $\mathcal{H}_L$, in which only the differences between the flows within each pair (i, j) of economic sectors are considered.

The RPB model is of great interest as it is related to both infinite-dimensional spin glass models, like the Sherrington-Kirkpatrick (SK) model [2], and to the Traveling Salesman Problem (TSP) [3,4]. The Hamiltonian of the SK model and related models is given by

$$\mathcal{H}_J(\sigma) = -\sum_{i<j} J(i, j) S_i S_j \tag{5}$$

with the configuration σ being comprised of N spins S_i, usually taking the values ± 1. Here J is a symmetric matrix with $J(i,j) = J(j,i)$ describing the interaction between the spins S_i and S_j. The optimization problem consists of finding optimum spin values leading to a minimum energy value. The traveling salesman has the task to find that permutation σ, for which the Hamiltonian

$$\mathcal{H}_J(\sigma) = J(\sigma(N), \sigma(1)) + \sum_{i=1}^{N-1} J(\sigma(i), \sigma(i+1)) \tag{6}$$

becomes minimal. Here $J(i,j)$ denotes the distance between two nodes i and j, usually measured in units of either length or time. Thus, the RPB model is right in the middle between the spin glass models and the TSP: like in the SK model, all entries of the upper triangular part of the exchange matrix are added to the cost function value, but one tries to find the optimum permutation for minimizing the energy like for the TSP.

For the TSP, Kobe and Klotz introduced a frustration measure m [5] which they called misfit parameter and which is defined as

$$m = \frac{\mathcal{H}_J(\sigma_0) - \mathcal{H}_J^{\text{id}}}{\mathcal{H}_J^{\text{id}}} \tag{7}$$

with σ_0 being the optimum configuration and $\mathcal{H}_J^{\text{id}}$ being the cost function of an idealized unfrustrated system in which each node i can be connected to its two nearest neighbors $n_1(i)$ and $n_2(i)$, i.e.,

$$\mathcal{H}_J^{\text{id}} = \frac{1}{2} \sum_{i=1}^{N} J(n_1(i), i) + J(i, n_2(i)). \tag{8}$$

The larger the misfit, the larger the deviation from this idealized trivial problem and thus the larger the frustration. A similar measure can be defined for the RPB problem as

$$m = \frac{\mathcal{H}_J(\sigma_0) - \mathcal{H}_K}{\mathcal{H}_K}, \tag{9}$$

with σ_0 being the optimum hierarchy and $\mathcal{H}_K$ being the constant energy value of the symmetrized trivial problem.

2   Other Ways of Defining a Hierarchical Order

The question arises why such a complex optimization problem has to be proposed for finding the hierarchical order of the various sectors of an economy and whether this problem definition could not be replaced by a much simpler approach, which introduces classification figures for all sectors, sorts these numbers according to their sizes, and leads to a result identical to the order one gets after an exact optimization of the RPB problem. There are various ways of how to define such classification figures, which shall be illustrated with a small, randomly created toy example. For this small instance, let $N = 4$ be the number of sectors and

$$J = \begin{pmatrix} 6\ 7\ 2\ 5 \\ 8\ 8\ 7\ 1 \\ 1\ 7\ 8\ 2 \\ 4\ 6\ 1\ 3 \end{pmatrix}$$

be the exchange matrix. When defining the classification figure of sector i as

$$c_i = \sum_{\substack{j=1 \\ j \neq i}}^{N} J(i, j), \tag{10}$$

we get the values $c_1 = 14$, $c_2 = 16$, $c_3 = 10$, and $c_4 = 11$. As $J(i, j)$ is the value of the products sold from sector i to sector j, c_i is the sum of the values of the products sold by sector i to all other sectors, such that this observable considers the problem of finding a hierarchical order from a supplier's point of view. When ordering the values of c_i according to their sizes, we get the hierarchical order $\sigma_c = (3412)$, as $\sigma_c(N)$ has to be the largest supplier. Alternatively, we can define a classification figure as

$$\tilde{c}_i = \sum_{\substack{j=1 \\ j \neq i}}^{N} J(j, i) \tag{11}$$

and thus consider this problem from a user's point of view. Here we get the values $\tilde{c}_1 = 13$, $\tilde{c}_2 = 20$, $\tilde{c}_3 = 10$, and $\tilde{c}_4 = 8$. We achieve the hierarchical order $\sigma_{\tilde{c}} = (2134)$, as $\sigma_{\tilde{c}}(1)$ has to be the largest user. The two observables c and $\tilde{c}$ consider different aspects of this problem and thus lead to different results. There is of course a way of combining these aspects by considering the differences between the entries just as in the last section and defining

$$\hat{c}_i = c_i - \tilde{c}_i = \sum_{j=1}^{N} (J(i, j) - J(j, i)). \tag{12}$$

Here we get the values $\hat{c}_1 = 1$, $\hat{c}_2 = -4$, $\hat{c}_3 = 0$, $\hat{c}_4 = 3$ and thus the order $\sigma_{\hat{c}} = (2314)$. When solving the RPB optimization problem for this toy instance, we get the unique optimum configuration $\sigma_{\text{RPB}} = (2431)$ with a ground state energy value of 22. The other configurations exhibit cost function values in the range $[23; 29]$.

Summarizing, we find that these different approaches lead to different hierarchical orders and that none of these simple approaches is able to produce a result which is optimum for the optimization problem proposed by Ramsey, Pierce, and Bowman. Although these simple approaches consider different aspects of a hierarchical ordering and might also have their justifications in some economic theories, we want to stick with the complex problem Ramsey, Pierce, and Bowman had to solve in their context, which intends to find a hierarchical ordering from the supplier's and the user's point of view simultaneously.

3   Computational Results for a Benchmark Instance

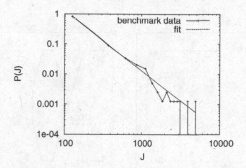

Fig. 1. The distribution function of the nondiagonal nonvanishing entries in the exchange matrix J of the benchmark instance is given as $\sim 1.3 \cdot 10^4 \times J^{-2}$.

In their publication of 1969 [1], Ramsey, Pierce, and Bowman also provided a benchmark instance comprised of a 37×37 matrix J based on the exchange of goods and services within the United States of America of the year 1947. Those entries in J that are marked as asterisks and thus classified as marginal in [1] shall be set to zero. The nonvanishing nondiagonal entries of J are power law distributed with an exponent of -2, as shown in Fig. 1. The mean value of all nondiagonal entries is ~ 108.75, the maximum is 4804. The authors were able to find an optimum solution for this benchmark instance consisting of 37 economic sectors with the ground state energy $\mathcal{H}_J(\sigma_0) = 25306$, consisting of $\mathcal{H}_K(\sigma_0) = 18650$ and $\mathcal{H}_L(\sigma_0) = 6656$.

For this publication, this benchmark instance shall be treated with the Simulated Annealing (SA) algorithm [6] and its deterministic variant [7], which is usually called Threshold Accepting (TA) [8]. When applying SA, the proposed optimization problem is considered as a classical physical system, which is gradually cooled down in an annealing process, thus being transferred from a high-energetic unordered regime to a low-energetic ordered solution. In each temperature step, several moves are applied to the system changing the configuration. When using SA, these moves are accepted according to the Metropolis acceptance criterion [9] with the acceptance probability

$$W(\sigma_{\text{current}} \to \sigma_{\text{new}}) = \begin{cases} 1 & \text{if } \Delta\mathcal{H}_J \leq 0 \\ \exp(-\Delta\mathcal{H}_J/T) & \text{otherwise} \end{cases} \tag{13}$$

with the energy difference $\Delta\mathcal{H}_J = \mathcal{H}_J(\sigma_{\text{new}}) - \mathcal{H}_J(\sigma_{\text{current}})$ between the current configuration σ_{current} and the tentative new configuration σ_{new}. T denotes the temperature. For TA, the acceptance criterion

$$W(\sigma_{\text{current}} \to \sigma_{\text{new}}) = \begin{cases} 1 & \text{if } \Delta\mathcal{H}_J \leq T \\ 0 & \text{otherwise} \end{cases} \tag{14}$$

is used. Thus, every move is accepted which leads either to an improvement or to a deterioration of a size which must not be larger than the threshold value T. In case of rejection, one sets $\sigma_{new} = \sigma_{current}$.

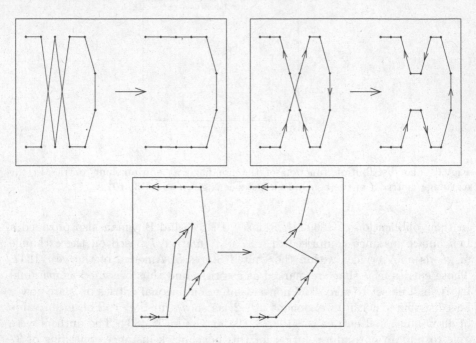

Fig. 2. Graphic illustration of the moves used for changing the configuration of the RPB model which is drawn like for a TSP with open end points: EXC (top left), L2O (top right), and L3O (bottom)

As the system size $N = 37$ of this benchmark instance is very small, it is sufficient to work with small moves only which do not change a configuration very much. Three small moves which are also commonly used for the TSP [10] have been implemented for the RPB model and are graphically illustrated in Fig. 2, in the way as if they were applied to a TSP with open end points. The various sectors of the economy are drawn as points on the plane. They are connected with edges picturing the sequence of these sectors in the current configuration. The Exchange move (EXC), which is shown in the top-left picture of Fig. 2, swaps two economic sectors in the configuration σ. The Lin-2-Opt move (L2O) [11,12] cuts two randomly selected edges in the sequence σ, turns the partial sequence between these two cuts around, thus changing its direction, and reconnects it with the usually two other partial sequences. Please note that in contrast to the TSP, for which mostly a closed tour is considered, the RPB model has open boundary conditions, such that also the cases have to be considered that there is no partial sequence before and after the partial sequence to be turned around and the whole sequence changes its direction after the application of the L2O, or that there is only one of them. The Lin-3-Opt move (L3O) cuts three randomly selected edges in the sequence, thus usually creating four partial sequences, exchanges the positions of the second and the third partial sequence, and reconnects the partial sequences. Again, there are special cases of the L3O, in which there are overall only two partial sequences to be exchanged or three partial sequences, two of which change their positions. Each of these three move routines is called with probability $\frac{1}{3}$.

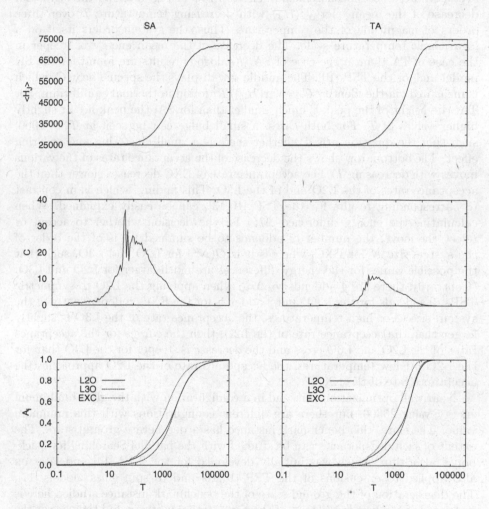

Fig. 3. Computational results for the application of SA (left) and TA (right) to the benchmark instance of the RPB problem: mean energy $\langle \mathcal{H}_J \rangle$ (top), specific heat C (middle), and acceptance rates A_i for $i \in \{$ L2O, L3O, EXC $\}$ (bottom)

Analogously to the SK model and the TSP [10], one finds that it is best to cool this benchmark instance of the RPB problem in an exponential way; a cooling factor of 0.9 is used. The initial temperature is determined automatically in a short random walk at the beginning, the final temperature is set to 5×10^{-3}. At the beginning of each temperature step, 10,000 Monte Carlo sweeps (MCS) are performed before the first measurement is taken. 5,000 measurements are taken, between which 10 MCS are performed. The results shown in Fig. 3 are averaged over these measurements. The top row of Fig. 3 shows the sigmoidal decrease of the mean energy $\langle \mathcal{H}_J \rangle$ with decreasing temperature T over three orders of magnitude of the temperature. Thus, the system orders itself on a logarithmic temperature scale. The decrease of the mean energy is steeper in the case of TA than in the case of SA. Analogous results are found for the SK model and for the TSP [10]. The middle row displays the specific heat C, which is measured via the identity $C = \mathrm{Var}(\mathcal{H}_J)/T^2$ found in thermal equilibrium. For TA, the height of the peak is much smaller than for SA, the peak lies at slightly higher values of T. For both cases, a small bulge can be seen at $T \sim 1000$, such that the question arises whether there is a small clustering and ordering effect. The bottom row shows the decrease of the acceptance rates of the various moves with decreasing T. The acceptance rate of EXC decreases slower than the acceptance rates of the L2O and of the L3O. This finding, which is in contrast to corresponding results for the TSP [10,13], can be easily explained. When calculating the energy difference $\Delta \mathcal{H}_J$ for the decision whether to accept or reject the move, the number of addends to be summed up is of the order of the system size N for EXC, whereas it is $\mathcal{O}(N^2)$ for L2O and L3O, such that the possible values for the energy differences are much larger for L2O and L3O. (Contrarily, there are 4 addends to $\Delta \mathcal{H}_J$ when applying the L2O to symmetric TSPs, 6 addends for the L3O, and 8 addends for the EXC, independently of the system size.) For high temperatures, the acceptance rate of the L3O is slightly larger than the acceptance rate of the L2O, then the curves for the acceptance rates of the L2O and L3O cross and the decrease is steeper for the L3O than for the L2O. At low temperatures, the acceptance rate of the L2O approaches the acceptance rate of the EXC.

Nearly all optimization runs end in a configuration with the global minimum energy value 25306, but there are different configurations with this minimum value. Therefore, this benchmark instance has a degenerate ground state. The extent of such degeneracies can be studied with the parallel Searching for Backbones algorithm, which was initially developed for the TSP [14] and later on also applied to extensions of the TSP [15,16] and to spin glass models [17]. The degeneration of the ground state of the benchmark instance studied here is restricted to the first five entries of the permutation vector σ and thus lies at the side of the largest users, whereas there is a strictly given optimum order for the largest suppliers. The largest users are non-profit organizations, amusements, scrap & miscellaneous industries, eating & drinking places, and ocean transportation, whereas the largest suppliers are transport via railroad and trucks, electric power plants, and banks and insurance companies. Please note that the results for the degeneracy and for the otherwise hierarchical order of the various economic sectors apply to this benchmark instance only. Mostly, the ground states of instances of the RPB problem will not be degenerate.

4   Conclusion and Outlook

In this paper, the properties of the widely unknown Ramsey-Pierce-Bowman problem are investigated: they introduced a model for finding a hierarchical order of the various sectors of an economy, in the viewpoint of users, which need to buy a lot of products from other sectors for their own production, and of suppliers, which earn money by providing their goods as preproducts to other sectors. According to different economic theories, either the largest suppliers or the largest users are the most important sectors of the economy. Therefore, it is of great necessity to know them. It turns out that this problem is located between the Traveling Salesman Problem, in which also an optimum sequence has to be found, and infinite-dimensional spin glass models, which make also use of the complete upper triangular part of the interaction matrix. After the derivation of some interesting properties of this model and showing that it cannot be trivially solved, the original benchmark instance is optimized using Simulated Annealing and the related Threshold Accepting optimization technique. It is found that the solution of this problem provided in [1] is indeed optimal, but that the benchmark instance exhibits a degenerate ground state. The computational results are similar to those obtained for the TSP and the SK model.

We will continue the investigation of this model, especially by comparing these results for data of the year 1947 with results for more recent years in order to see the change of the importance of the various sectors for the US economy. We expect that e.g. the scrap industry has changed its role from a user to an important supplier. The RPB model has also many other applications: in the foreign trade, suppliers have a trade deficit, whereas the users have a trade surplus. In political sciences, the movement of voters between the various parties is studied, here the suppliers are those parties which lose votes to other parties. A further application is the investigation of migration processes of peoples: users are immigration countries, suppliers emigration countries. Furthermore, we want to investigate the properties of the exchange matrices of these and related problems and see whether they also exhibit scale free properties like the exchange matrix of the benchmark instance we investigated here. Finally, we also want to study the problem of detecting the importance of an economic sector as user or supplier also with other algorithms, e.g. with a flow analysis using the infomap algorithm [19].

Acknowledgment. JJS and AL would like to thank Gunter Dueck and the members of his former TopC optimization group at the IBM Scientific Center Heidelberg, especially Tobias Scheuer, Hermann Stamm-Wilbrandt, and Gerhard Schrimpf, for fruitful discussions about breaking world records of benchmark problems (see also [18]). We also kindly acknowledge Fred Ramsey (Department of Statistics, Oregon State University) offering a prize for beating his solution for the benchmark instance we considered in this paper. However, we were only able to produce equally good solutions. All computational evidence leads us to conclude that no better solution for this benchmark instance exists. Furthermore, JJS is grateful to Dietrich Stauffer for fruitful suggestions and humorous remarks. JJS would like to thank the John von Neumann Institute for Computing at the Research Center Julich, Germany, for a generous grant of computing time.

References

1. Ramsey, F.L., Pierce, D.A., and Bowman, V.J.: Triangularization of input-output matrices. Technical Report No. 16, Department of Statistics, Oregon State University, 1969; Proceedings of the Business and Economic Statistics Section of the American Statistical Association, vol. 676 (1969)
2. Sherrington, D., Kirkpatrick, S.: Solvable model of a spin-glass. Phys. Rev. Lett. **35**, 1792–1796 (1975)
3. Lawler, E.L., Lenstra, J.K., Rinnoy Kan, A.H.G., Shmoys, D.B.: The Traveling Salesman Problem. John Wiley and Sons, New York (1985)
4. Reinelt, Gerhard: The Traveling Salesman. LNCS, vol. 840. Springer, Heidelberg (1994). https://doi.org/10.1007/3-540-48661-5
5. Kobe, S., Klotz, T.: Frustration: How it can be measured. Phys. Rev. E **52**, 5660–5663 (1995). and references therein
6. Kirkpatrick, S., Gelatt, C.D., Jr., Vecchi, M.P.: Optimization by simulated annealing. Science **220**, 671–680 (1983)
7. Moscato, P., Fontanari, J.F.: Stochastic versus deterministic update in simulated annealing. Phys. Lett. A **146**, 204–208 (1990)
8. Dueck, G., Scheuer, T.: Threshold accepting: A general purpose optimization algorithm appearing superior to simulated annealing. J. Comp. Phys. **90**, 161–175 (1990)
9. Metropolis, N., Rosenbluth, A.W., Rosenbluth, M.N., Teller, A.H., Teller, E.: Equation of state calculations by fast computing machines. J. Chem. Phys. **21**, 1087–1092 (1953)
10. Schneider, J.J., Kirkpatrick, S.: Stochastic Optimization. Springer, Berlin (2006)
11. Lin, S.: Computer solutions of the traveling salesman problem. Bell Syst. Tech. J. **44**, 2245–2269 (1965)
12. Lin, S., Kernighan, B.W.: An effective heuristic algorithm for the traveling-salesman problem. Oper. Res. **21**, 498–516 (1973)
13. Stattenberger, G., Dankesreiter, M., Baumgartner, F., Schneider, J.J.: On the neighborhood structure of the traveling salesman problem generated by local search moves. J. Stat. Phys. **129**, 623–648 (2007)
14. Schneider, J., Froschhammer, C., Morgenstern, I., Husslein, T., Singer, J.M.: Searching for backbones - An efficient parallel algorithm for the traveling salesman problem. Comp. Phys. Comm. **96**, 173–188 (1996)
15. Schneider, J., Britze, J., Ebersbach, A., Morgenstern, I., Puchta, M.: Optimization of production planning problems - A case study for assembly lines. Int. J. Mod. Phys. C **11**, 949–972 (2000)
16. Schneider, J.: Searching for backbones - A high-performance parallel algorithm for solving combinatorial optimization problems. Future Gen. Comput. Syst. **19**, 121–131 (2003)

17. Schneider, J. J.: Searching for backbones - An efficient parallel algorithm for finding groundstates in spin glass models. In: Tokuyama, M., Oppenheim, I.: 3rd International Symposium on Slow Dynamics in Complex Systems, Sendai, Japan, 3–8 November 2003, AIP Conference Proceedings, vol. 708, pp. 426–429 (2004)
18. Schrimpf, G., Schneider, J., Stamm-Wilbrandt, II., Dueck, G.: Record breaking optimization results - Using the ruin & recreate principle. J. Comp. Phys. **159**, 139–171 (1999)
19. Rosvall, M., Bergstrom, C.T.: Maps of random walks on complex networks reveal community structure. PNAS **105**, 1118–1123 (2008)

Ethics in Computational Modelling

Ethical Aspects of Computational Modelling in Science, Decision Support and Communication

Rudolf Marcel Füchslin[1,2]([✉]), Jacques Ambühl[3], Alessia Faggian[4],
Harold M. Fellermann[5], Dandolo Flumini[1], Armando Geller[6], Martin M. Hanczyc[4],
Andreas Klinkert[7], Pius Krütli[8], Hans-Georg Matuttis[9], Thomas Ott[10],
Stephan Scheidegger[1], Gary Bruno Schmid[11], Timo Smieszek[12],
Johannes J. Schneider[1], Albert Steiner[7], and Mathias S. Weyland[1]

[1] Institute of Applied Mathematics and Physics, Zurich University of Applied Sciences Zurich,
Technikumstrasse 9, 8400 Winterthur, Switzerland
furu@zhaw.ch
[2] European Centre for Living Technology, Ca' Bottacin, Dorsoduro 3911 Calle Crosera,
30123 Venice, Italy
[3] World Bank Group, Geneva, Switzerland
[4] Laboratory for Artificial Biology, Department of Cellular, Computational and Integrative
Biology (CIBIO), University of Trento, 38123 Trento, Italy
[5] School of Computing,
Newcastle University, Science Square 1, Newcastle Upon Tyne NE4 5TG, UK
[6] Scensei (Switzerland) GmbH, Fraumünsterstrasse 11, 8001 Zurich, Switzerland
[7] Institute of Data Analysis and Process Design, Zurich University of Applied Sciences,
Rosenstrasse 3, 8400 Winterthur, Switzerland
[8] Transdisciplinarity Lab, ETH Zurich, 8092 Zurich, Switzerland
[9] Department of Mechanical Engineering and Intelligent Systems, The University of
Electrocommunications, Tokyo 182-8585, Japan
[10] Institute of Computational Life Sciences, Zurich University of Applied Sciences, Schloss 1,
8820 Wädenswil, Switzerland
[11] Swiss NeuroCreations GmbH, Hambergersteig 25, 8008 Zurich, Switzerland
[12] Muesmattweg 70, 4123 Allschwil, Switzerland

Abstract. The development of data science, the increase of computational power,
the availability of the internet infrastructure for data exchange and the urgency
for an understanding of complex systems require a responsible and ethical use of
computational models in science, communication and decision-making. Starting
with a discussion of the width of different purposes of computational models,
we first investigate the process of model construction as an interplay of theory
and experimentation. We emphasise the different aspects of the tension between
model variables and experimentally measurable observables. The resolution of
this tension is a prerequisite for the responsible use of models and an instrumental
part of using models in the scientific processes. We then discuss the impact of
models and the responsibility that results from the fact that models support and

© The Author(s) 2022, corrected publication 2023
J. J. Schneider et al. (Eds.): WIVACE 2021, CCIS 1722, pp. 263–293, 2022.
https://doi.org/10.1007/978-3-031-23929-8_24

may also guide experimentation. Further, we investigate the difference between computational modelling in an interdisciplinary science project and computational models as tools in transdisciplinary decision support.

We regard the communication of model structures and modelling results as essential; however, this communication cannot happen in a technical manner, but model structures and modelling results must be translated into a "narrative." We discuss the role of concepts from disciplines such as literary theory, communication science, and cultural studies and the potential gains that a broader approach can obtain. Considering concepts from the liberal arts, we conclude that there is, besides the responsibility of the model author, also a responsibility of the user/reader of the modelling results.

Keywords: Ethics · Computational modelling · Transdisciplinarity

1   Introduction

This article deals with the responsible and ethical use of computational models in science, communication and decision-making. We intend to report experiences we collected over the last couple of years and a focussed discussion during a workshop at the WIVACE 2021 conference, held in Winterthur, Switzerland from Sep 15 to Sep 17, 2021[1]. A systematic treatment is given e.g. in [1] or earlier [2]; for a recent discussion based on case studies see [3], and an overview with a focus on philosophy see [4]. The societal role of artificial life has been a central topic at the ALIFE 2019 conference in Newcastle, UK; various articles on the topic can be found in [5]. Computational modelling still increases its importance in science and, as we have seen during the COVID-19 pandemics, is recognised as a central supportive tool in political decision–making processes. We, as modellers, realised the necessity for embedding and relating our work into a framework that also includes ethical considerations. In this work, we report on our findings, mainly derived not from theoretical work but our daily practice.

This report has been made possible by the joint efforts of different research projects. The leading role is thereby with the EU-funded project ACDC (Artificial Cells with Distributed Cores to Decipher Protein Function, funded by the European Union's Horizon 2020 research and innovation programme, https://acdch2020.eu/). The workshop was organised in an open format but initiated and supported by ACDC. We deliberately have chosen to collect examples from different fields to achieve generality and not restrict ourselves to specific research areas. Together with the open nature of the workshop at WIVACE 2021, this fact justifies the rather long list of co-authors. The first author takes the principal responsibility for the paper, the contributing co – authors are listed in alphabetical order. An extended version of this paper will be the base of a report to the EU as a deliverable for the ACDC project.

This article focuses on computational models and not on theory in general. In our view, computational models are a subclass of general models. A general model may define the terms of discourse; a computational model adds quantifiable relations between these terms (or, as we will define it later, uses variables subject to computation). These

---

[1] https://www.wivace2021.org/.

quantifiable relations are the basis for the implementation of a simulation. A further note on computational models and machine learning: We wrote this report with a somewhat "physics-oriented" interpretation of the term "computational model." Roughly said, we discuss models in which the model variables have an interpretation/ a semantics right from the start and do not acquire this interpretation throughout a training phase. We compare this to machine learning, say, an artificial neural net. The weights attributed to connections of neurons in a neural net have no interpretation before the network's training. Furthermore, even after the training, relating these weights to observables after the training phase is a complicated and only partially understood task. However, we point out that almost everything we write in this article applies to models with variables with direct semantics as well as structures such as artificial neural networks. We do not discuss the issue of the semantics of internal variables further, but we emphasise its importance and point out that it has many layers. For an eloquently written discussion of the "symbol grounding" – problem, i.e. the question of how internal variables are related to objects in the internal world, see [6]. Note, for example, that symbol grounding can become very intricate if one deals with a model that works with variables with a probabilistic interpretation.

Computational models have always been of importance. However, confluent trends of the last couple of years have increased the role of models in science and the relation between science and society. It is a sign of hope and the reason for optimism that recently, powerful youth movements and responsible politicians recommended and even demanded to unite behind the sciences. As scientists, we should welcome this trust. However, we are obligated to reflect on how we can justify this trust and how we have to communicate the results of computational models to avoid misunderstandings and prohibit misuse.

The responsible and ethical use of models is the main topic of this report. We discuss diffrent aspects of the question of responsibility and ethics. Responsible use is closely related to several recurring requests R1-R3:

- R1: A clarification of the differences between computational models as tools in science and computational models that are part of decision support processes that go far beyond the social context of science itself.
- R2: A better understanding of interface processes. Models and simulations are powerful instruments for linking different fields of human expertise and establishing a relation between the real world and different abstractions of it. First, this linking between reality and abstraction requires the construction of interfaces. The presence of interfaces usually implies some form of translation processes, which in general leads to systematic information loss (because a model only represents a part of reality) as well as different types of translation errors (which, for example, can be the result of the limited precision of measurements). Second, models can help to connect different (abstract) universes of discourse (to use the term from computer science) or languages specific to social groups. As scientists, we are used to working within some more or less well-defined area of discourse in which a common language exists, and the discussion participants share an implicit understanding of boundary conditions, interpretation rules, and the like. As soon as one uses the results of scientific discourse as input for a more general and interdisciplinary decision support process, one must

not take this implicit understanding for granted. Scientists have to prepare the output of a computational model in a way that is digestible by actors outside the field in which the model has been set up.

- R3: Not only those developing and using computational models in order to produce output have a responsibility. There is also a responsibility of the reader. By the term "reader," we mean those who take up the result of a computational model but may not have area-specific knowledge to interpret these results as it is standard by those who are within the area of expertise. The scientists have to require the reader to be aware of (probably area-specific) limitations of computational models but also have to explain these limitations in a form that the reader can understand.

In the communication with the public and members from scientific fields not directly connected to computational modelling or natural science, it became clear that (at least) two main issues need to be addressed: the purpose of models and the relation of theory and data science.

The first issue (purpose of models) is a very fundamental one. Many people believe that a model is always a tool for setting up predictive simulations. We summarise this idea of a computational model in Fig. 1. The figure illustrates the model-based control of a robot as a predictive tool for controlling the robot's dynamics. The critical aspect is that the simulation produces a sufficiently faithful analogue of the dynamics in the real world in an appropriately designed mathematical representation. The term "sufficiently" refers to the quality of the control of the robot. Somewhat loosely expressed, sufficient means that the control ensures that the robot reaches a given set of goals. One can draw a similar picture of a simulation for weather forecasts. Also, the simulation result should give a sufficiently accurate prediction of reality in that context.

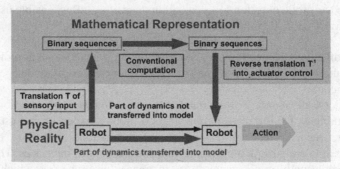

Fig. 1. Computational models as tool for the control of a robot. The computational model serves as a tool for predicting the dynamics of a technical system in a non-trivial environment.

Prediction is an obvious goal of a simulation but certainly not the only one. A non – exhaustive list of (in parts overlapping) purposes for models includes:

- Check the **understanding** of the past: By comparing model output and measured data, one can check whether the model explains what has happened or whether the model is insufficient and must be extended.

- **Optimise** the present: This applies to static or stationary situations with parameters. If one has an optimisation goal, *in silico* experiments can help find optimal values for these parameters.
- **Predict the future** if one knows the present and has a reliable model of the dynamics: Ideally, a model allows one to predict the future, given sufficient knowledge of the initial conditions. However, even if the initial conditions are known, the prediction of the future may still be imprecise, partly because models most often only approximate the actual dynamics, partly because many models contain stochastic components. We point out that even if the latter is not the case, i.e., the dynamics of the model is entirely deterministic, the future may still be difficult to predict, mainly if the dynamics of the model exhibits chaotic components.
- **Estimate the future** if one only guesses the present. Most often, one has only limited knowledge about initial conditions. In that case, either one estimates the initial conditions by statistical means (including the values of the parameters determining the dynamics) or constructs initial conditions based on plausible assumptions.
- **Explain what we see:** A model can help to give meaning to data[2] in the sense that the dynamics or a state can be explained by referring to the model's variables (see below the discussion of Fig. 2). We point out that the purpose of "giving meaning to data" is probably the one in which a restriction of the discussion to computational models is least necessary. Models as tools for explanation do not necessarily need to produce computational results but can serve as platforms for defining concepts and interactions qualitatively. Furthermore, note that an explanation of the behavior of variables (e.g. by showing correlations) differs from the first item in the list, which refers to the question whether the model can reproduce what has been observed.
- **Analyse the dynamics**: it is a fact that human beings are pretty good at understanding equilibrium states but are quite often surprised by the consequences of feedback processes, second-round effects, stochasticity (the consequences of fluctuations), or non-linear dynamics in general. This is an ideal application for models and simulations: with the help of simulations, one can get "a feeling" for the dynamics of a model and explore relevant settings by trying out the consequences of changes in inputs, variations in parameters and the like.
- Detect **emergent dynamics and structures:** Quite often, systems exhibit emergent phenomena. There are many examples in which a model sheds light on the underlying processes by showing that a specific mechanism leads to an observed emergent phenomenon. However, we point out that a model may well give a plausible explanation for certain emergent phenomena; this plausibility must not be confused with evidence.
- **Explain system behaviour from more fundamental dynamics**. A model may explain the behaviour of a specific system from the fundamental properties of its constituents. In engineering, a material model (i.e. a model taking material constants and fundamental physical laws as input) may explain a whole family of systems, such as FEM – models in civil engineering or the study of biological matter using molecular dynamics simulations.

---

[2] This has been pointed out by Marcello Pellilo from the University Ca'Foscari, Venice.

- **Decision Support:** Even if, because of lack of data about initial conditions, a model cannot give sufficiently reliable information about the future development of the system under consideration, it may still help to evaluate the range of possibilities via scenario analysis. Such an analysis contains reasonable best and worst cases and a quantitative evaluation of the sensitivity of parameters. Especially studies of parameter sensitivity may help allocate potentially limited resources to determine those parameters that influence and output in a critical manner.
- **Models as platforms for discussion:** Models can serve as platforms for discussions (between modellers and experts or even between experts with domain knowledge with the modellers as moderators) of assumptions, parameter dependencies and qualitative aspects of system behaviour.
- **Virtualisation:** A sufficiently precise model of reality may enable "in silico" experimentation. Besides benefits concerning costs and speed, virtual experiments enable the study of seemingly unphysical situations, e.g., by turning off selected physical mechanisms. Such "knockout experiments" shed light on the actual physical circumstances' importance.
- **Produce data and train modern controllers** (e.g., deep neural nets): various types of artificial intelligence and, more generally, machine learning are now part of the modeler's toolbox. Many of these algorithms require vast amounts of data for training, often much more than experiments can provide. One can train neural nets with simulated data, comparable to the training of pilots in a flight simulator.
- ...

The second main issue relates to the role of data. In 2008, Chris Anderson wrote an article in Wired titled "The End of Theory: The Data Deluge Makes the Scientific Method Obsolete.", [7]. Although the article's content was much more nuanced, the upcoming of data science and the broad availability of data brought some people to conclude that models and theories are unnecessary. We agree that the availability of cheap sensors, the possibility to transfer data from the sensors to some data processing unit without the need to install complicated hardware, and the ease with which the vast amounts of data can be analysed changes science in a deep sense. However, we still think that models are of relevance. We point out that the amount of data needed to replace a model or theory is often underestimated (particularly in the social sciences). The need for large amounts of data means that even if it were possible in a fundamental sense to dispense with models and only rely on data, it may still not be practical. In addition, one may discuss whether data can replace models in all circumstances. If one used computational approaches only for prediction or scenario building, something like the "Master Algorithm" envisioned, e.g., in [8], could, in principle, do the job. However, other model purposes, such as giving meaning to data, are hard to conceive without a model based on variables with semantics. We also point out that data-driven approaches usually perform poorly in generalisations, at least in the present situation. One may well predict a specific system's behaviour using a data-driven approach, but this ability for prediction is usually challenging to transfer to a different system. In contrast, a system modelling platform based on a fundamental material model relating variables with physical interpretation predicts the behaviour of large classes of structures and systems. The tension between purely data–driven approaches and semantics lies at the heart of the discussion about "explainable AI", see e.g. [9] and with emphasis on medicine, [10].

It goes without saying that in this article, we do not claim to give final answers to the request R1-R3; our experiences over the last couple of years in various disciplines enable us to report some of our findings and some generalisations distilled out of these findings over the last couple of months. Thereby we span modelling of evolutionary processes, statistical physics, modelling of cellular processes, model-based therapy optimisation, traffic simulations, applications of AI in medicine and control of large industrial entities, to give some examples.

We point out the importance of the last two years. All the contributors to this article have worked for decades in various applied and fundamental sciences fields, always using or developing computational models. In addition, some of us are now serving in the decision support for the Swiss government concerning the Covid 19 – pandemics. The change from publishing results in peer-reviewed journals to more direct decision support gave us (sometimes for the first time) the experience of science that has a short-term impact. However, it also made us reflect on the responsibility of developing and using computational models. In addition, it became clear that there is a considerable difference between scientific modelling for science itself and the use of scientific models as tools in a broader, nonscientific (or not exclusively scientific) but rational context.

This report does not focus on the necessity of quality control and standard operating procedures but more on aspects of communication. Also of crucial importance is that scientists being part of decision support processes should reflect on their role. Our experiences led to establishing a network of scientists and decision-makers discussing the role of computational models based on our recent research activities, ranging from purely scientific activities in EU-funded projects to decision-support. Our goal is to study and describe the difficulties of computational modelling in a broader context and in a permanent manner that includes publications and network activities.

The article is organised as follows: Sec. 2 discusses our perspective on models, model building, and implementation. The section defines a couple of terms and presents our view of the process of model building. Sec. 3 discusses the use of models in science, which exhibits relevant differences to the use of models in decision–making. The latter involves a full transdisciplinary mode of communication and is treated in Sec. 4. The interaction of science with stakeholders outside the scientific discourse requires the use of according means of communication (which we call narratives). In Sec. 5, we postulate a responsibility on the narrator's side as well as on the side of the reader. We emphasise the potential for science and science communication to learn from the vast body of literary theory, for the practice of communication, but also its conceptualisation. In the conclusions (Sec. 6), we relate our findings to the current status of modelling in society.

2   The Process of Modelling

There is a broad discussion of what exactly one understands by the term "model." This discussion ranges from literary theory over philosophy to the foundations of mathematics and logic (where mathematical structures such as the natural numbers serve as "models" for sets of axioms, e.g., the Peano axioms). In order to reduce misunderstandings that may occur if one speaks about a concept that appears in many different branches of science, we start with defining some basic terminology that we will use in this article:

- A **model** establishes relations between different classes of objects (which we will call fundamental terms). These relations can be static but also consist of rules and descriptions of the dynamics of these objects. Importantly, we require a model to be based on a rational description of the objects and their interactions. Thereby, we understand by the term "rational description" a language-based, sufficiently intersubjective formulation that a sufficiently knowledgeable group of experts can understand.
- A **theory** describes the fundamental terms and interactions of the model in a context that may well go beyond the scope of the model. To give an example here: relativistic quantum field theory is a theory in the sense of a framework, and the standard model of particle physics is a model that one builds within the framework of this theory.
- A **fundamental term** is an object of consideration in its broadest sense. In other words: the fundamental terms of the model are the objects the model is dealing with and talking about.
- A **variable** in a model is a numerical representation of such a fundamental term. We use the term variable in a loose sense as it can be either a single number, a list of numbers, or an instance of a more complicated class object.
- A **parameter** is a number that characterizes some aspects of the interaction or processing of the variables. The difference between parameters and variables may depend on context and is often somewhat arbitrary.
- A **computational model** is a model expressed in variables that can be subject to data processing.
- An **observable** in an experiment is a quantity for which a feasible measurement process exists.
- A **representation** of reality (or briefly representation) is a relation between observables and variables of the model. Ideally, there is a one-to-one correspondence between variables in a model and observables in an experiment. In any case, there should be some relation between observables and model variables in order to constitute a relationship between the model and the reality.
- An **implementation** of a model is some software that translates a computational model in operations on some hardware.
- If we execute an implementation of a model with some input variables, we call this run a **simulation**.

The focus on computational models has at least two direct consequences:

1. Since we usually work in an interdisciplinary context, we have to use software implementations and data formats that are widespread and easy to handle. In practice, we are restricted to data that can be expressed in the form of real or integer numbers. If the fundamental terms of the model are not generically numbers (e.g., in the case of sentiment analysis or image analysis), we must be aware that we need a mapping from the non-numerical terms to some sort of numerical representation. Such a mapping comes with its difficulties and is usually the source of various errors, some

of which are systematic. Although the development of artificial intelligence probably will enable the classification of more complex data, e.g., the analysis of graphs, these methods are not yet widely available or easy to use outside of relatively narrow contexts. The range of problems one encounters if one deals with non-numerical data is broad: It starts with the fact that the translation of non-numerical into numerical data usually requires some classification. The classification criteria themselves are often chosen in an ad hoc manner and do not rely on a clear scientific strategy. The criteria reflect expert knowledge but lack proper rationalisation.

2. If the first restriction results from the request that the fundamental terms of the model are expressable as numbers (or sets of numbers), the second is about the simulations we perform with these numbers. Simulations are only helpful if one can efficiently do them. That means algorithms and hardware must allow performing the necessary computation within a timescale that is compatible with the needs of those who take the model outputs bases for decision making.

Based on these considerations, one must be aware that the construction and implementation of the model is a process that requires several tightly connected steps. Figure 2 gives an illustration of the process as we see it.

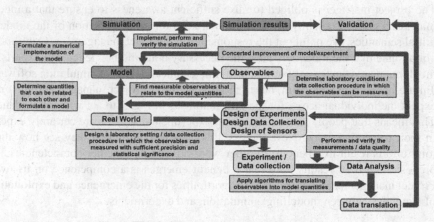

Fig. 2. The process of modelling.

Our analysis of the model building process establishes the first set of responsibilities: one has to perform all indicated steps according to the standards defined by the area of expertise and the methods one uses. These standards include scientific aspects and standard operating procedures in managing complex processes such as software development. We will not discuss this first set of standards in detail here. They are, again depending on the field, developed to a high level of sophistication, sometimes even formulated in terms of norms. Instead, we focus on two aspects that seem to be obvious but are nevertheless not intensively discussed and, in our experience, the source of problems in many projects that include computational models, namely modelling as a systemic task and the difference between implicit and explicit knowledge.

2.1 Modelling is a Systemic Task

The most important lesson one has to learn in modelling is that a model is not an entity independent of the user and the experimental input (by the term "user," we understand those taking the results from a simulation but not necessarily involved in the technical aspects of model building). From the project management point of view, it turns out to be a significant challenge to orchestrate the interaction between users, data providers/experimentalists, and those developing and implementing a model:

- The users must be well informed about what they can expect from a model and what is outside its scope. Whereas a model may be suitable for, say, the purpose of formulating scenarios or making the qualitative aspects of the system's dynamics transparent, it may not be able to produce reliable predictions. Different reasons may cause this: for example, the input data cannot be given with sufficiently high precision. Alternatively, the model itself bases on assumptions that may not be appropriate in the situation under consideration. Responsible project management must clarify what the different partners can expect from each other.
- The modellers often have a solution that seeks a problem (in practice: one aims to transfer software developed for the simulation of one type of system to another one). The project manager is obliged to raise sufficient awareness to ensure that a model matches a given problem without too much need for reinterpretation of the model's initial semantics. We point out the responsibility of project management: One should not assume that the modellers have the necessary domain knowledge and experts in the domain are usually not familiar with the technical details of simulation software.
- Taking up the previous point, but in more generality: Project management must not expect the individual actors for the different tasks in Fig. 2 to act without guidance. This means that people in simulation and experimental science are certainly experts in their respective fields. Nevertheless, this does not imply that they see how their joint efforts result in a benefit. In other words: Bringing together competencies and orchestrating their results such that a benefit emerges is a competence on its own. Project management must recognise opportunities for the emergence and exploitation of synergies between modelling, simulation, and experiments.

A specific but frequent challenge emerges at the boundary of pure science and application. In pure science, one often works with so-called "toy models." Their purpose is to elucidate the mathematical structure of the model and to study qualitative features of the emerging dynamics of a given system. Sometimes, such models are surprisingly powerful, despite their deliberate simplicity (see, e.g., the Ising-model in the study of phase transitions of magnetic systems. There is a temptation to transfer such models from a qualitative into a quantitative, real-world context. A well-known example that illustrates the situation in our view quite well is Schelling's segregation model. The Schelling model shows how micro-motives can transparently lead to macro behaviour. Although one can learn an important lesson from this model and its simulations, it is certainly not a tool that one should use in urban planning, and it is certainly too simple to represent and explain actual social dynamics; for a very illustrative discussion see [11].

2.2   Modelling Requires Turning Implicit into Explicit Knowledge

What one can model and simulate depends on the knowledge that can be injected into the model-building process in Fig. 2. In what follows, we distinguish implicit and explicit knowledge. By the term explicit knowledge, we understand the knowledge that

- is well-defined and can be expressed in some reasonably general language such that it can be communicated to sufficiently well-informed non-experts outside the domain of expertise on the consideration,
- is sufficiently formalised (quantitative and qualitative) that one can translate it into some algorithm.

In comparison, implicit knowledge consists of unwritten but (within the domain of expertise) generally accepted assumptions and standard operating procedures. In the context of science, this difference is quite often well understood, in the sense that we may know from experiments that a particular procedure works but not always why this is the case. However, even if we lack this knowledge, we are in science usually aware of this fact, and various practices have been established to deal with this situation (e.g., phenomenological models and explanations).

The situation is different if one uses models for decision support. Many historically grown social and economic structures are not well understood, even though they are working in a very stable manner (one may argue that it is precisely that stability that made a detailed analysis unnecessary). To give a famous example of a seemingly simple consumer good: shoelaces (we learned about this example in a televised interview with the German economist Hans Werner Sinn, who brought it up to illustrate a similar point). It is probably possible to get shoelaces in almost all locations worldwide, provided a certain standard of living has been achieved. It is reasonable to assume that there is probably no single human being who understands all aspects of the production and distribution of shoelaces in all detail. Nevertheless, the supply of shoelaces seems to be a self-organised process without any plan (or any underlying model). Shoelaces are a simple product; today's technology and politics give examples on all levels of sophistication (the information flow in a medium-sized enterprise usually differs from what one reads on the organigram, and the resilience of supply chains in a globalised economy is a very recent and intricate topic). In the situations described, one is not even capable of formulating a phenomenological model; not only does one not know how the system's components interact, but the components themselves also may not be known. Whereas experimentalists in a scientific project may not know everything about the system they analyse, they usually have a good understanding of what they do not know. In contrast, the actors in a social system may not even know that the system exists.

Self-organisation is ubiquitous in human societies, and deliberately so, at least in liberal ones. At first glance, it may look reasonable to analyse socioeconomic processes in sufficient detail such that a model of that process can be set up. In a very optimistic worldview, turning implicit into explicit knowledge is a process that is worth it on its own. We adhere to this position in the context of science but point out some issues that one needs to consider, particularly if one applies models in a broader context:

- A model may shift established balances of power. If a predictive model existed of a sufficiently large part of society, those who have access to this model would have considerable power. Modelling enables control, and control implies power, which can be misused.
- Applied to socio-economical and socio-technological systems, a model (particularly but not exclusively a successful one) seduces to centralise process control. This may be sensible, especially for cost-effectiveness, but may hinder the self-organisation of processes. Self-organised processes have specific advantages (e.g., usually they are resilient), which is not always the case for centralised processes. One has to evaluate carefully whether one wants to give up the benefits of self-organisation, and the first step on this path is the construction of a model.

The consequence of these considerations is by no means a request for less modelling. Nevertheless, if models result in power, we should enforce and guarantee transparency and equal or, at least, democratically controlled access to the models; for the importance of transparency, see, e.g. [1]. This holds for models guiding political processes and, as well, models in science (the term "democratic" then refers to the scientists involved in the project and the standards and practices of their respective fields).

2.3   Models Can Enable Thinking but May also Set up Limitations

In science, the problem of models that are (mis-)used for exercising power seems to be of minor relevance. This is because, in science, the role of models is usually not the control of processes but insight. Nevertheless, if a computational model turns out to be supportive of experiments, a sort of feedback may start to take effect. In practice, models are not complete. And even if they were, there would undoubtedly be settings that are easier to simulate than others for technical reasons. Then, there is a tendency to do what can be calculated and not necessarily what is most interesting from a pure domain-specific perspective. This may be perfectly reasonable, but one should bear in mind that the question of whether a process can be simulated is somewhat extrinsic to the process itself.

This aspect becomes apparent in teaching, where models play an important role. One often discusses simplifications and idealisations such as frictionless movements or perfect crystals. One tends to choose examples because they result in equations that can be solved with the mathematical means available to the students in the context of the applied model. But: Whether or not a process is easy to calculate does not necessarily say something about its relevance to nature; this is a deep issue, see [12]. Restricting the analysis of processes to that one can compute may be reasonable, especially in teaching. However, one should always keep in mind that there are phenomena outside the somewhat artificial boundaries imposed by requesting computable models.

The fact that models may impose limitations on scientific investigations is not restricted to teaching. As an example, we point out that the notion of an "integrable system" in the theory of dynamical systems is a fundamental concept, somewhat loosely defined as systems with conserved quantities and therefore restricted to submanifolds of the phase space under consideration. Chaotic behaviour is closely related to integrability, or better, its absence. Even though it was already known in the 19th century that the behaviour of dynamical systems could be very complicated and unstable, in more general science, the notion of "chaos" came up only around 1960, and somewhat as a surprise to scientists used to integrable models. Still today, at least in engineering, chaotic behaviour is often regarded as the exception (and not, as it is, in fact, the rule).

As a side remark, engineering is an interesting case in that respect. One could claim that a large part of engineering consists of the attempt to construct systems so that they can be described and controlled by efficiently computable models.

2.4   Interdisciplinary Work is No Excuse for Diffusion of Responsibility

A more general issue concerns the convergence of experiments and models. The different tasks in Fig. 2 require several different fields of expertise. In practice, one observes the danger of a certain diffusion of responsibilities. This problem is well-known in a single scientific project and may be avoided by project management. As soon as one starts to use models as tools for decision support and works in a broader setting, the problem of the diffusion of responsibilities becomes somewhat more pronounced.

On the one hand, modellers tend to complain about the "lack of data" and use this argument to justify the shortcomings of models. Project management must clarify that the limited availability of data may look like a bug but is actually a feature to deal with responsibly. That means: one must set up models in a manner that can get along with the available data.

On the other hand, there is a specific danger that those using the output of models do that by regarding the models as black boxes and putting trust in them without scepticism. The scepticism certainly includes the results produced by the model. A responsible and practical form of scepticism is constructing a broad range of plausibility checks. Those who use the model's output usually have a rather precise idea of what this output should look like, especially for some extreme choices of system parameters. Such checks can be pretty efficient, but one needs to organise them properly. Again, in the relatively narrow setting of a scientific research project, this practice is well established and can be implemented easily (probably because experimentalists are familiar with checking their setup by measuring some boundary cases). As soon as one uses models for general decision support, the employment of plausibility checks by extreme scenarios requires some management skills. Besides the fact that the communication between those implementing a model and those applying the results has to be organised (and is subject to limited resources, especially time), a psychological barrier has to be passed. Project management must clarify that models usually do not work right from the start but require certain debugging and polishing. We repeatedly encountered considerable criticism from those using the results of models at the initial stage of implementation. The line of argument was: "If your model cannot deal with a straightforward situation, how can we trust its results in a more complicated setting?". Responsible project management makes it transparent at an early stage that an iterative calibration and refinement process is an instrumental part of modelling.

According to our experience, there is a potential misunderstanding about modelling: Sometimes, users are under the impression that modelling means implementing a small number of very fundamental relations and natural laws, which, if implemented correctly, will result in precise predictions in all possible settings of the input variables. There is a hidden danger in this optimistic perspective. If a model were omnipotent (at least within the frame of the system simulated) framework for prediction, the task of posing appropriate questions would be trivial. Say it in other words: if the modellers can simulate everything, the users can ask anything. However, models are most often much more specialised in the sense that the range of questions they can answer reasonably is limited. For the design of a model and a simulation, it is essential to know what questions the users want to answer. Responsible use of models requires carefully designing questions and being aware of the limitations of models. This holds especially if models are developed in a collaborative process of modellers and users. It is then the responsibility of the users to define in sufficient detail what types of results the model should be able to produce.

We illustrate this with examples from science and decision support. Within a specific scientific discipline, for example, solid-state physics, the relation between experimental observables, the model's variables, and the difference between qualitative and detailed quantitative statements are well understood by all partners. Discussions may still be necessary, but they build upon a tradition that is part of the discipline (see also Sec. 3.2). Again, the prototypic example is the Ising-model; see for example [13]. Interdisciplinary discussions are more demanding than intradisciplinary ones. A famous example is the application of Lotka-Volterra models on problems in ecology, e.g. the lynx – snow hare – cycle. A mathematical model can never predict in detail how populations develop, and it is also not possible to determine their exact size in the field. The question is then what one can learn from such model, i.e. what questions are answered by the model and what conclusions would be an overstatement. The discussion between modellers and ecologists should then focus on whether and how the model's output can be made helpful for ecological reasoning. Such discussions become even more necessary in a transdisciplinary context (s. Sec 4.1). We refer to epidemiology, where coarse-grained models get input data of somewhat limited quality (the data reported may be satisfactory for, say, influenza but do not match the needs for a novel disease because essential aspects are not analysed). Note well that there is a trade-off between the resolution of the model and the amount of input data necessary. A finer resolution would require even better data, see also the next section. For decision support during an ongoing pandemic, the modellers have to clarify that this trade-off exists and that predictions about, for example, the number of fatalities cannot be made (although giving bands is possible). In contrast, statements about, for example, the transition into an endemic state are possible with reasonable reliability.

We emphasise the importance of this discussion. In decision support, the scientists and modellers must realise that their partners from, say, politics are not necessarily familiar with the do's and do not's of a specific scientific discipline. Nevertheless, these partners have to communicate science-based decisions. In our view, modellers must actively support this communication and in a way that reflects the specific challenges of communication with a general audience.

2.5   Observables are not Variables

It may sound relatively trivial, but it turns out to be a fundamental management challenge to bring model variables and experimentally measurable observables into a relation. Or, to state it differently: in a project that uses models, variables and observables must be chosen to be related efficiently and require tools available to those working on the project. This matching process is complicated.

This complication results, first, from intrinsic reasons. Model users and modellers are not necessarily aware that variables and observables are linked by a process that includes several steps (for a still simplified overview, see Fig. 3). Here, $S(t)$ represents the system under consideration at the time t. We point out that by S, we don't understand a number or some other data structure but an actual system. The downward track represents experimentation. Some aspects of the physical system can be observed; these are the observables $O(t)$, which are still understood as physical phenomena. The relation/mapping between system and observables is given by a relation $O(t) = \Theta(S(t))$. The observables can be measured, which produces signals $M(t) = \Gamma(O(t))$. The signals are assumed as data structures, e.g., time series of numbers or digital images. We point out that, for example, the production of a digital image is again a process that involves many steps, but most of them are standardised and/or their limitations are purely technical and do not involve the issues we are raising here. Going upwards from S represents the modelling track. First, the real world has to be mapped to a mathematical model $R(t) = \Phi(S(t))$. $R(t)$ are mathematical objects of some kind. For obtaining a computable model, these mathematical objects need to be represented by some data structure $N(t)$ (which, at the end, is a finite bit string): $N(t) = \Psi(R(t))$. The data structures $N(t)$ are the model

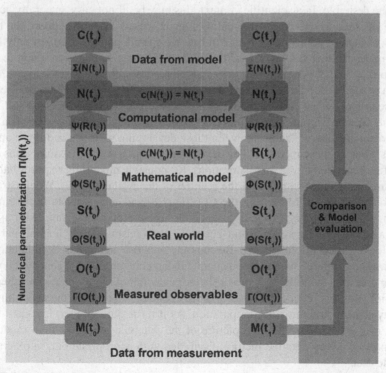

Fig. 3. Relating observables and variables. After comparison and model evaluation, the model and the experiments may be modified, s. Fig. 2.

variables. We point out that this step requires some subtle considerations: The mathematical objects $R(t)$ need not to be conventional numbers but can be complex numbers, vector fields, manifolds, or other mathematical objects. The translation of, for example, a vector field into a data structure is a step that poses its specific challenges (the choice of a coordinate system, for example). Finally, model variables, by some processing, are transformed into data $C(t) = \Sigma(N(t))$, which one can compare (in a mathematical, means quantifiable sense) with the signals $M(t)$. The possibility for this comparison is instrumental for model validation and is the basis for model parameterisation.

Second, modellers and experimentalists have somewhat different objectives. The more precisely one can describe an object, the better it is for the model, or to be precise, the easier modelling is, at least from a conceptual perspective. Relying on probabilistic concepts or statistical quantities such as averages is done chiefly if enforced by limited computational power or memory space. However, data acquisition by experiment requires specialised equipment, time, and staff. It sometimes is even impossible (from a modelling perspective, in vivo data acquisition would be desirable for calibrating models of cellular processes, but the necessary experimental possibilities are still not always available).

The task of responsible project management consists of finding a position that matches given experimental boundary conditions. This means setting up a permanent negotiation process between modellers and experimentalists oriented on the possible and not on the desirable. In a scientific context, this may be, if not easy, generally accepted practice. As soon as one enters the realm of model-based decision support, the tension between available and desirable data can become considerable, with often only limited mutual understanding for, e.g., legal boundary conditions (to mention a non-technical limitation).

As a general observation, we note that model building tends to be output-oriented in the sense that modellers want to produce optimal results but often do not care sufficiently about the necessary input. In part, this is a reasonable practice because there is a historically grown division of labour between those developing methods and those applying the tools based on these methods. The developers tend to work with assumptions or "toy data" often generated based on assumptions. Responsible modelling requires balancing the desire for maximal output with the realities of the available input. We illustrate this with a recent example. Modelling the COVID -19 pandemics was an important challenge in the last two years. Modelling pandemics is a prototypical example for applying agent-based models. However, agent-based models constitute only a small part of all simulations in epidemiology [14]. Besides the fact that agent-based simulations are rather time-consuming, one must consider a fundamental problem. Agent-based models allow, in principle, to model very precisely the behaviour of representative batches of the population. But this precision comes at a price: one has to provide the according input data, say contact structures in a population. As it turns out, simpler, less data-hungry models may give a more reliable picture of the potential scenarios than models that would provide a very detailed picture if only fed with appropriate but, in practice, not available data.

3 Models in Science

There is a vast literature about the role of models in science; for a short and well-written overview, see, for example [15]. In our practice, two categorisations are of particular relevance for ethical considerations. First, one can use models for pre- and post-processing of data. The two uses pose different challenges concerning responsibility. Second, one can classify computational models in those trying to virtualise a given situation as precisely as possible (we call these models complete) and minimal models. The focus of minimal models is generality; one looks for general mechanisms and strips them from the details of specific systems. One explores those aspects of the dynamics independent of the details of this implementation.

3.1 Pre- and Post-processing Data

One can roughly categorise the role of computational models in science into two classes: Pre- and post-processing data of an experiment (whereby pre-processing includes those cases where one does not perform an experiment at all).

Post-processing happens in cases where one pursues one or several of the purposes discussed in the introduction. "Understanding the past" and "Giving meaning to data" are undoubtedly important cases, whereby meaning includes the transformation of sensory signals into measurements, s. Figure 3, but also includes the support of the interpretation of data.

Pre-processing experimental data encompasses virtual testing, scenario building, and prediction. For example, microscopic models can relate fundamental properties of a system to some macroscopic observables. From a scientific perspective, such microscopic models are crucial for the reductionist program. For engineering, material models enabling virtual construction are vital tools in modern design processes.

From an ethical perspective, the distinction between post- and pre-processing is relevant insofar as in post-processing, the further use of outputs of an experiment stays in focus, whereas pre-processing may guide the implementation of the experiment itself. One could think that, because the experiment already happened, post-processing data is of minor ethical relevance. This is, of course, not true. For example, in most industrialised countries, the regulations for animal testing require ethical balancing, which means a justification of animal suffering compared to human benefit (see e.g. European directive 2010/63/EU[3]). The use of models can do both, enhance the benefit of data from an experiment, and change the type or reduce the amount of experimentation for the same benefit.

Somewhat colloquially said, the challenge of pre-processing lies in the fact that models start to guide experiments. Using models as guidance can be misleading in various ways: A system may seem to work in theory or *in silico*, but does not in reality (It is only slightly polemic to say that this is not a real problem but what experimentalists expect from modellers). Another danger is that a model may show something to be unfeasible, which some clever engineering can implement nevertheless. Even if we assume that the modelling has been done with all precautions established in the field or

---

[3] https://eur-lex.europa.eu/eli/dir/2010/63/2019-06-26.

reasonably possible: All modelling is based on assumptions that are sometimes by no means obvious, and one tends to disregard them. In our experience, only a permanent and maximally transparent discussion of assumptions between all involved parties in a project can reduce the problem of (mis-) guidance. Again, we point out that this process is by no means easy and an actual intellectual task (One needs to pass the boundaries between different fields of expertise).

3.2   Minimal and Complete Models

Again, we discriminate models into two large classes, whereby we are aware that a continuous spectrum would be more appropriate. By a minimal model, we understand a model that is as simple as possible and contains only those variables of primary interest. In a minimal model, one deliberately simplifies external effects, environmental conditions, and complicated details of the interaction between the parts of the system under consideration. Such models turned out to be of enormous value to science. As examples, we note the Ising model in statistical physics or the Schelling model in the study of segregation processes (by the way, the mathematics behind these two models are closely related). Minimal models can shed light on qualitative aspects of the system's dynamics and, despite their seeming simplicity, can show rich emergent behaviour. However, the purpose and scope of the model needs to be clarified as soon as minimal models are used outside a narrow range of scientific investigation. Results produced by minimal models are often easy to interpret, mainly because the relation between cause and effect is transparent. The problem with minimal models is that it is difficult to transfer their results into real-world situations. What seems to be a plausible cause and effect relationship in a minimal model needs not be one in the real world.

Complete models try to represent as much of reality as is possible or necessary. Put costs aside, this seems to be a reasonable approach. However, such models contain a lot of parameters and turn out to be difficult to calibrate. The main difficulty of calibration is that if there are various parameters, their respective values may be underdetermined regarding the available data. At this point, one usually quotes John von Neumann, who is said to have said: "With four parameters I can fit an elephant, and with five, I can make him wiggle his trunk". Today, partially caused by machine learning development, science developed several sophisticated procedures to cope with underdetermination. In our view, the problem with complete models is that the preparation of input becomes a highly complex task. This is first because the input tends to be large. Second, some types of input data are easier to get than others in reality, which leads to biases. Systems in which one couples different processes, some that one can parameterise by laboratory experiments and some that cannot be isolated (e.g. socio-technological systems, for which the technological processes are usually easier to parameterise than the social components) pose a particular problem. The more input parameters one has, the more difficult it becomes to evaluate the quality of the output because whether or not the limited precision of an input parameter has a relevant effect depends on the size of the statistical uncertainty and the influence of that parameter. Robustness analysis is often only of limited help because biases in input parameters, mainly if one deals with social systems, tend to be correlated.

We emphasise that phenomena observed in minimal models are generic in that they are not the result of some, potentially very special, circumstances of a specific situation. From this perspective, minimal models are general models.

4 Models in Decision Support

Models have a well-established and continually refined role in science, with a development based on practical experiences and theoretical studies. Especially with the growing digitalisation but also the awareness for data science, models get an essential role in decision support, be it in an emergency such as COVID-19 pandemics or be it in planning, e.g., in an economic or urban context.

One may argue that decision support is somewhat outside the boundaries of science, and one should not mix up scientific and nonscientific applications of models. However, we advocate that the distinction between scientific and nonscientific uses is somewhat artificial and neither beneficial for the goals of science nor modelling as a scientific field. First, the boundary between science and non-science is challenging to draw. In our experience, the discussion of what belongs to science rarely leads to a sensible conclusion and most often ends in a somewhat sterile dispute about definitions.

We see a considerable benefit in the broader use of models. First of all, we believe that models can contribute to decision-making. In addition, if there is bidirectional communication between users and modellers, confronting a model with reality can promote science by adding novel questions and initiating developments that have the advantage of dealing with testable scenarios. Often, external questions can promote interesting internal scientific developments. If at all, the distinction between pure and applied science can be characterised by the notion that pure science studies the internal workings of the system under deliberately simplified boundary conditions (the laboratory conditions), whereas applied science asks for what can be said scientifically about a system immersed into a complex real-world environment. Note well that in this reading, applied science is not just about applying the results of science. Applied science also aspires to do science outside the laboratory.

Understood in the manner described above, the authors are all involved in applied sciences, particularly science applied to decision support. This includes applications of modelling in purely scientific projects such as ACDC, in which a sophisticated interaction between modellers and experimentalists is a key project objective. This involvement also includes decision support in processes that include the whole society.

We investigate two large clusters of problems, the first one relating to the position of science in society and the second addressing some hidden aspects of optimisation. We conclude with some general observations, which we realised as being important.

4.1 Transdisciplinarity

We distinguish inter- and transdisciplinarity by assuming interdisciplinarity as communication and collaboration over the borders of different branches of sciences but within the general context of science. We understand transdisciplinarity as the interaction between the actors inside and outside science. For a detailed discussion of these concepts, see [16] or [17].

Science is a cultural practice with its own rules, language, codes of conduct, and signaling systems. As a scientist, one must accept that communication with the nonscientific sphere requires finding common grounds and using a common language. There are several challenges, which we will address in what follows.

Arrogance: Science and Democracy

Science is undoubtedly one of the most successful collection of practices in human history. This fact is seducing. Scientists tend to justify the advantages of a scientific mindset and approach by the successes of science. Some conclude that because science works well, not only those doing science but everybody should act and think like a scientist. There is a danger of going even further and regarding those unfamiliar with what scientists call a scientific discourse as unfit for a general discussion.

We regard this as a naïve position. Although we are thoroughly convinced of the value of science and its methods, we are aware that either the concept of a "scientific discourse" must be stretched until it contains all forms of rational discourses or essential questions, e.g. about moral values or inter-subjective impressions as they occur in art, are not regarded as part of the discussion. For computational modelling, this means that even if one is convinced to have powerful tools that enable a well-grounded insight into natural and social processes, these insights are not above democratic processes. The scientists must make these insights a part of democratic negotiations that respect, for example, discussions about moral values. These discussions may still be rational but are not necessarily following what one usually regards as integral parts of the scientific method (evidence-based argumentation, falsifiability, etc.).

Analogies, Metaphors and Speaking Plain English

Transdisciplinarity requires abstaining from jargon. We point out that this requires a lot of effort. In fact, "jargon" in science means the use of sophisticated formalised concepts, which in some cases require a level of abstraction that one can only master after intensive occupation with a specific topic and its formal apparatus. We emphasise the role of formalisation; often, the underlying idea of a concept is well accessible; the formalisation of the concept requires training and detailed knowledge about the formalisation itself (for a detailed discussion of this point in the context of the physical sciences, see [18].

There is a real problem: One cannot explain complex formal abstractions "in a sentence." We see only three, partly connected, ways out of this:

1. Speaking in analogies: One compares something unfamiliar with something more familiar to the audience. We emphasise that there are various aspects to consider. First, when explaining models, analogies most often focus on functions, structures and dynamic behaviour. Second, they are rarely faithful. That can be helpful because they enable highlighting relevant aspects and neglect those of lesser or no relevance. On the downside, one always has to keep in mind the danger of overstraining analogies. Third, if A is an analogy of a process P, the audience may be familiar with A's formalisation but not with the one of P. This is a situation of particular interest

because one can scrutinise the extent to which two formalisms are equivalent with some rigour, and one can state the limits of analogies. At the same time, one can profit from the power of formalised reasoning.

2. Explanation: One tries to explain formalisations. As desirable as it would be, this is often unfeasible.

3. Modularisation: One modularises the explanation of a complex process and makes the modules, their dynamics, and their interactions transparent. Thereby, the processes taking place on the level of modules should be clear to all participants of a discussion but not necessarily the internal dynamics of the individual modules. An example is a recipe: The act of cooking requires the realisation of a series of biochemical processes. The (evolved) practice of cooking modularised the process by using building blocks that are robust (small deviations of temperature, amount of ingredients and the like lead to only small changes in the outcome; processes close to transitions such as caramelisation are for the advanced), standard cooking ware is employed, and there is usually no need to know about the chemistry or physics that takes place.

We emphasise two further issues. First, the relation between concepts and formalisation is not unidirectional. The formalisation itself may lead to an extension of the initial concepts. A famous example is antimatter (or, to be precise, the anti-electron, now termed positron), which P. A. M. Dirac introduced on purely mathematical grounds. Such concepts, which originate not from experience or experiment but as a consequence of mathematical reasoning, are particularly hard to convey to an audience without formal background.

Second, we point out the difference between analogies and metaphors. We quote H. U. Fuchs: "We all have access to abstract schemas that form through organism-environment interactions. Understanding something (or making something understandable) means bringing a description/explanation back to these fundamental abstractions/schemas (which are used in metaphors, and metaphorical webs, which, in turn, are used in narratives).", see also [19]. In order to explain processes/structures, one uses metaphors; by analogies, one compares and relates processes/structures.

A narrowed version of a metaphor is the idea of an abstract or conceptual data structure, say a vector. Vectors are inspired by the combined notion of length and direction. The concept of a vector is embedded in a network of other concepts, such as angles, rotations, parallelograms, dimensions, et cetera. Furthermore, our intuitive notion is complemented by a rigorous mathematical formalisation (which, in turn, inspired further concepts such as infinite-dimensional vectors for which we lack a complete intuition, see above). We discuss this example because it highlights a significant problem. The representation of a vector in a model (the respective model variable as it appears in Fig. 2) is usually an array of numbers. For communication, it is essential to understand the difference between the representation of a variable and the concept behind it. The representation carries much technical baggage, such as coordinate systems, which bury the idea of a vector.

In a communication, using the representation instead of the concept may be tempting for those familiar with the former; but it is usually not helpful for those lacking this technical familiarity. Where does the temptation come from? Speaking in plain English about representations is usually relatively easy (which is why we can "explain" representations even to computers using, from a linguistic perspective structurally simple, programming languages), whereas verbalising concepts requires hard work and accurate language skills.

Questions and Answers in Science

Much postmodern critique of science tries to show that science is a social construct, and therefore, the scientific method has no privileged position for understanding the world. We cannot enter this discussion on a broader level; concerning computational models, we have to address some questions and points of critique:

1. As natural scientists, we take the existence of an "objective reality" as a fundamental assumption. However, one must not misunderstand a model for this reality. It is at least a point to keep in mind that a computational model is defined and bounded by many constraints (some are economical and therefore structurally social).
2. From the point of view of a natural scientist, the answers of science are at least approximations to objective truths; but the according questions are not. If one regards the development of science and particularly modelling in a transdisciplinary context as an interplay between questions, answers, refined questions, and consequently the further development of methods, complex models do have socially constructed aspects.
3. A computational model needs input data. As discussed in Sect. 3.1, the input data can be subject to ethical evaluations. Since what one can compute is a function of the available input data, the ethical considerations concerning input data affect the possible modelling results.

Even if one does not share postmodern positions, the points above show that computational modelling, especially if applied in a context that includes partners from fields of expertise outside of science, is certainly affected by social and cultural processes.

4.2 Optimality: Give Options, Not Advice!

Decision support often aims to find an optimal strategy or implementation of procedures for a given task. One major challenge if one works outside a strictly scientific context is finding a proper definition of optimality. The result of any attempt to find an optimal solution depends on what one regards as desirable. As shown in what follows, the discussion of how one defines optimality has various aspects and lies in our view at the core of the ethical aspects of modelling. The problem is multi-layered: Even if one has a quantifiable desirable goal, there may still be several additional boundary conditions that one has to observe to establish procedures that yield optimal results and do this in a fair manner.

Optimality and Fairness

In a purely technical context, optimality is quite often easy to define. Even then, one must be aware that proper balancing may not be trivial if there are different criteria for optimality (for example, efficiency and efficacy or quality and output in engineering). This is also recognised in a business context [20], where the discussion about KPIs (key performance indicators) has reached a high level of sophistication.

In a political context, optimality is most often a question about values. It is a hallmark of democracy that such questions have no general and definite answer (derived from some dogmatic set of principles) but are subject to permanent discussion in each case.

Optimality becomes even more involved if resources are limited, and one includes criteria considering the fairness of distribution. Here, computational models can be beneficial. As an example, we mention [21], a study in which the distribution of a limited number of defibrillators over different areas has been investigated. At first glance, optimality is easy to define and given by the number of saved lives. However, this would imply placing the defibrillators preferentially in urban areas, where many potential patients can profit from their presence. A fair distribution should not disadvantage those living in rural areas. In [21], a computational model was used to find a distribution pattern that maximises the number of saved lives and considers a fair distribution of resources. From the perspective of computational modelling, one has to quantify and combine two different criteria for optimality. The quantification makes implicit valuations explicit (compare with Sec. 2.2), and this itself poses non-trivial political challenges.

Ethics and Second-Round Effects

The distribution pattern of defibrillators is an example of a static situation. Computational models are advantageous if they clarify ethical considerations in dynamic processes. Here, one has to distinguish between first-round effects and subsequent processes, which we summarise here by the term second-round effects (being aware that there are third-, and in general and nth-round effects). We illustrate this by a study [22] that used models to optimise the distribution of limited vaccines in influenza pandemics. A general goal is undoubtedly to maximise the number of saved lives. In the case of influenza, this implies that in general (there are exceptions, though) that the efforts should be focused on the most vulnerable. There are two different ways how the most vulnerable can be protected. First, they can get a vaccination. Second, they can be protected from infection by reducing the number of contacts with already infected ones. If one studies a "common influenza" and does not consider measures such as lockdowns or quarantines (which in 2017 looked outlandish), one can reduce the spread of the disease by vaccinating that part of the population first, which contributes most to the distribution of the infection. In general, the group of the most vulnerable (in the case of influenza, usually the elderly) and the group of the most critical spreaders are not identical. Whether direct vaccination (the first-round effects) or protection by reducing the spread of the disease and vaccinating the spreaders first (a second-round effect) results in a maximal number of saved lives depends on various parameters and can be studied by a computational model.

As we realised, the communication of such second-round effects is far from easy. We point out that whether one directly vaccinates the most vulnerable or protects them by stopping the spread of the disease always serves the same goal, namely the maximisation of the number of saved lives (which means the lives of the most vulnerable). However, one must carefully explain why protection of the most vulnerable sometimes may be most effectively achieved by the prioritised vaccination of a different part of the population.

That second-round effects occur can often be made clear with qualitative arguments. However, whether or not they can become prominent, even dominant, is usually a quantitative question. A computational model helps to show first that there are settings in which second-round effects are relevant and second, which factors influence the extent of this relevance.

Optimality and Observables
In physics, a central principle states that the laws describing a process must be independent of the choice of coordinates one uses (That is an occasion where a subtle difference between models and simulations becomes apparent: one can formulate the physical laws describing the trajectory of an asteroid without referring to a specific set of coordinates. A simulation processes numbers and relies heavily on the choice of coordinates). Whereas one can firmly establish this principle in the natural sciences on mathematical grounds, the situation is much more complicated in socio-technical investigations.

We explain this with an example. In order to account at least partially for the variability of society, the individual members of the population are grouped into cohorts. Especially in medical settings, this usually happens according to age and medical preconditions. The interactions in the population are then also formulated based on these cohorts. One could use a different grouping, leading to different interaction schemes. Different groupings may represent the variability of society differently for the phenomenon under investigation and with different statistical quality. Whether age or, say, socioeconomic status is the best descriptor in a given situation is not always apparent, and the choice of the descriptors may influence the outcome of computation and the conclusions one draws from it. We emphasise that computational models that serve as tools for determining optimal solutions under consideration of ethical principles must be scrutinised for their dependence on the choice of input and model variables.

One may now ask for a determination of the best way to describe the process as a prerequisite for any use of computational models for ethical purposes. As reasonable as this sounds, it is pretty often not feasible. The evaluation of socio-economical data is difficult and expensive. As a modeller in decision support, one may be confronted with the fact that one has to work with the available data, which is not necessarily the data that would be best suited. It is, in our view, the central responsibility of the modeller always to point out that fact.

4.3 The Role of Experts

In inter- or transdisciplinary projects, experts from different fields have to interact. Groups of experts, especially modellers, should carefully reflect their roles as soon as they become part of the decision-making process. We identified two main issues that we address in what follows.

Groupthink
As pointed out in the introduction, one reason for using models is that they may help understand non-linear behaviour, emergent dynamics, and sometimes the appearance of seemingly counterintuitive phenomena. The challenge is distinguishing between those phenomena that are hard to understand but are real and those resulting from some possibly

wrong assumption underlying one of the various aspects of the modelling process. We repeatedly observed an interesting process that belongs to the class of problems that one usually summarises by groupthink. The process works like this: a model gives a hard-to-understand result. The modellers, usually not experts in experimentation or observation, ask those familiar with the experimental aspects of the system on the consideration of whether or not the result of the model can be true. The experimentalists, not being familiar with the internal workings of the model, give a plausible explanation for what has been computed with the simulation belonging to the model. Such a line of argumentation that relies on incomplete knowledge may be the source of groupthink. Each member of the group regards his/her partial knowledge as justified by the partial knowledge of other experts. The problem thereby is not so much that all involved parties only have partial knowledge; this is unavoidable in interdisciplinary work. The problem is: What seems to be plausible in a discussion should be based on evidence or more detailed scrutiny.

The Position of Experts in Complex Decision Processes

In our view, the most critical problem of the use of models in decision support is the necessity of the experts to develop a proper understanding of their role in the decision-making process. Especially when models are used to evaluate or optimise ethical aspects in the decision process, experts tend to advise decision-makers. This advice is usually based on an already preselected set of simulation results. In our view, modellers must avoid this preselection in a proper decision-making process. The experts, especially the modellers, should understand their role as giving options for decision-makers. These options are then used to achieve a proper decision and represent a range of possible further actions. The modelling results should show the consequences and costs of different potential courses of action but should avoid guiding the decision in a specific direction by imposing a value system that is not transparent to the other parties in the decision process.

The other side of this is the potential tendency of decision-makers to diffuse responsibility by taking the results of modelling as such and not to apply a prioritisation or a valuation based on a transparent and ethically grounded evaluation scheme.

To say it in one sentence: For maintaining the integrity and transparency of decision making, science gives options, and decision-makers value and select them.

4.4   Computational Models as Tools for Discussion

In 2.2, we pointed out that models require turning implicit into explicit knowledge. This process is necessary for model building but is also of use in decision support. The discussion of concrete model assumptions and the possibility of studying their influence at least in a semi-quantitative way (for example, whether specific output variables are positively or negatively coupled with some basic assumptions?) helps to understand the emergent mechanisms in complex processes. The discussion between the "users" with expertise and domain knowledge and the modellers who perform simulations (always in the context of a given model) can result in a better understanding of the system as a whole and the emergent properties one observes (in reality and the model). In an ideal case, the interplay of experts (scientists or non – scientists) who discuss their experiences and verbal descriptions with modellers who turn these statements into formalised algorithms can improve the levels of insight into complex systems.

A further benefit is that an algorithmic description can be communicated in a way that is sometimes hard to achieve by prose. This communication is not only crucial for the interplay between modellers and other stakeholders; in our experience, the discussion of a model and its algorithms can also be beneficial for the collaboration of different stakeholders. We then use the model as a tool for discussion, and modellers act as intermediaries.

This function as a tool also applies to situations where one discusses qualitative aspects of system behaviour. For example, we take the question of tipping points, i.e. a qualitative change of system behaviour resulting from a small change in one or several parameters. A minimal model (s. Sec. 3.2) can help decide whether some generic dynamical properties are sufficient to produce such a tipping point. Knowing about such a tipping point is of value, even if we know that the numerical value at which the transition happens in a minimal model may differ considerably from the one in a specific and complete setting. This type of discussion is well known in physics (we again refer to the Ising – model, which is a minimal model of magnetism but shows some behaviours of phase transitions generically).

5   Models and Narratives

We focus the discussion on a central topic: The interpretation of results gained from models happens in a series of steps. This interpretation starts in the context of science, the place of production. Various methodologies, "best practices," and cultural habits exist in this relatively narrow social environment. Later, various instances transfer these results into a language that suits broader, even public communication needs. The transfer is not a translation; transfer is not only a matter of using "plain language." Instead, the communicator produces a "narrative" in which common analogies replace the system or process under consideration (see Sec. 4.1.2). We claim, however, that the scaffolds of narratives appear at an earlier stage in the process of model building.

We start with a central hypothesis, from which we derive/on which we base several questions. We cannot answer these questions in a definite manner, but we recognised them as central for discussing the relation between models and narratives.

5.1   Main Hypothesis on Models, Simulations, Storyworlds and Narratives

Although models are based on quantitative or qualitative scientific reasoning, how they are perceived and used in a context broader than that of science should be analysed with a range of tools from communication, journalism, literary analysis, and critics. In the narrow context of science, a model is a basis for mathematical reasoning. The function of a model is broadened, as soon as models and their results become part of the thinking and acting in politics, administration, and the wider public. Besides mathematical arguments, the model sets the stage for narrative elements. Again, taking up an idea of H. U. Fuchs, the model gives a story world, and the simulation is the backbone of a concrete story.

In what follows, we will use the term "narrative" instead of "story". According to Collins dictionary, a narrative is "a story or an account of a series of events", whereas a story is defined as "a description of imaginary people and events, which is written or told in order to entertain".

In such a broad setting, "Reading the results of a model" is a non–trivial process that can no longer rely on some scientific subdiscipline's established standards and rules. Understanding the role of models/story-worlds and simulations/narratives and putting it into a social and ethical context requires a discussion that raises awareness of the reader's role and his/her background.

Good literature is more real than reality in the sense that a well-composed narrative contains "reality" in a more condensed and easier-to-follow form than just an account of what has happened where and when to whom. We probably all agree that writing a good narrative is a significant task. As soon as simulations are related to narratives, we strongly emphasise that communication profits from the inclusion of experts. However, we point out that one needs more than marketing (marketing is needed, but not only). One needs narrators and experts from literary studies who understand the complex relations between texts and readers.

5.2 Questions and Topics Relating to Politics and Operationalisation

Models and Novels
We compare a good model to a good storyworld: The model sets the stage for a narrative that, in some respect, is a streamlined image of reality but contains, concerning a specific set of topics, a sufficient representation of reality. Like a good novel, this narrative focuses on those parts of the dynamics relevant to the phenomena under consideration and neglects the others. As already explained in Sec. 4.2.3, socio-medical models often subdivide the population according to age. That leads to a picture in which "the elderly," "boomers," and "the younger" appear as actors. This is often reasonable but sometimes hides the fact that a similar subdivision according to socioeconomic status could be employed, which leads (literally) to a different narrative.

When one equips a model with a narrative, one needs to ask about the opportunity costs of invoking one specific narrative: The choice of the narrative one tells automatically implies that other narratives remain untold. Choosing a narrative (which happens already at an early stage in model-building when one chooses the model variables, see Fig. 2) must be done considering potential uses for ethical purposes later. Conversely, if one has a model and evaluates its use for ethical issues, one must ask whether the model is an appropriate stage for narratives that illustrate the ethical question under consideration.

Models, Narratives, and Communication Structures
In larger projects or organisations, models/storyworlds and simulations/narratives are embedded in communication structures. If one aspires to establish a smooth and correct interpretation of simulations in an organisation, a central question is: How can we avoid (maybe interest guided but probably more often unconscious) misinterpretations? The

problem of misinterpretation is closely related to the "Give options, not advice!" – statement we discussed in Sec. 4.3.2. The narrative should inspire thinking and discussion but not predestine their outcome not justified by the facts.

It is not only about misinterpretation but about interpretation in general. Note well: As soon as one accepts that models and simulations go together with narratives, we have to accept generic properties of the latter as a part of the whole process. To express this colloquially: The fact that narratives can have many interpretations is not a bug but a feature of literature.

One further point is that how a narrative is understood depends on the culture in which that reading happens. If narratives transport/communicate the results of models, we should compose narratives with an awareness of the difficulty of writing stories in an intercultural context.

Secondary Literature

One usually values the primary texts higher than the secondary sources in literature and philosophy. This is different in the natural sciences, where almost nobody learns, for example, quantum mechanics via reading the original papers. This may be a pity in some specific cases where the original works are written by true masters of the field and contain deep insights. In general, however, the secondary literature clarifies basic concepts and uses a more accessible presentation and improved formalisation. Secondary literature in the natural sciences is quite often easier to understand and, therefore, more efficient in teaching sometimes rather technical ideas. One can explain this observation partly by considering that the authors of secondary literature have been in the same situation as the novice is when studying a new topic: One has to master an idea that one has not produced by oneself. We emphasise this, because, in our view, one must not regard the communicator as solely supportive. Those translating a model into a narrative contribute an essential part of knowledge in a transdisciplinary process.

5.3   The Role of the Reader

One big difference between narratives in literature and narratives derived from scientific model-based simulations is the multiple authorship of the latter. Literary works with more than one author (not to speak of five, ten, or twenty) are almost non-existent and, if at all, are certainly part of the experimental branch of literature. Concerning ethical considerations, this raises important questions. What is the individual author's responsibility for the narrative produced from the simulation results?

One can extend this question. If we compare the narratives related to models and simulations with other literary works, is there an ethics of literature that helps us understand how we can deal with models/simulations? On the one hand, there is the freedom of artistic work (which is an essential aspect of art). On the other hand, narratives/literature have a social effect. This means that the concept of a narrative may originate and primarily exist in the realm of art but can, as an intellectual object, be used outside art and, consequently, be subject to different standards than in the context of art. The question is then: Can we apply the methods of literary criticism to narratives (an object initially in art) but use it outside art?

Through the investigations of philosophers such as Roland Barthes, we know about the reader's role. One may or may not share the positions expressed in [23]; taking the reader's role seriously means that there is a responsibility of the author(s) and one of the reader. From a (or at least some) modern point of view, the text exists for its own, and the reader may well go beyond what the author had in mind. Whereas this "going beyond original intentions" - approach is most fruitful in, say, reading poetry, it is more problematic in model interpretation. If modelling results are embedded into narratives, the storytellers and the reader must be aware that what they read is a narrative, but the interpretation is not as free as in the case of a pure work of art. Whereas it is appropriate to take a piece of art as inspiration for own thoughts, ideas and emotions, a narrative for simulation results must be regarded as a vehicle of content. In reading such a narration, the reader is required to scrutinize her or his interpretations and try reading the text in the author's sense. The authors have the obligation to make this sense transparent.

The presentation of the results of a simulation in a scientific manner (means as tables and graphs) has its advantages insofar as the potential for misunderstandings is reduced. If modelling and simulation results are embedded into narratives, the reader or user of the simulation results shares the responsibility for correct reading and interpretation. That means, for example, that users are responsible for knowing the difference between a scenario and a prediction. On a somewhat higher level, users must be aware that the quality of the input data determines the quality of the output data. In general, users of modelling results must understand their role not only as a receiving one but as, in various aspects, a critical part of modelling ("critical" in the sense of "important", but also in the sense of offering critique to the modellers). This holds for individual readers but even more so for the media.

6 Concluding Remarks

In the introduction, we formulated three requests for responsible modelling. In the paper, we focussed the discussion on computational modelling.

The first request addresses the users of models. Doing science and acting in the scientific community requires acquaintance with and acceptance of a specific set of social practices. If one acts in an interdisciplinary context but still within science, one can build on these practices. As soon as one enters decision-support, the involved partner may add different boundary conditions or potentialities that alter the extent and range of ethical considerations. The discussions in Sec. 2–5 contribute all to this discussion.

The second request, R2, postulates that one needs a process-oriented understanding of modelling. This holds for the development of models, discussed in Sec. 2. The interplay between model builders and experimentalists poses challenges for project management. In our view, it is crucial to understand that the existence of these challenges is not "a bug, but a feature." Of course, experimentalists and modellers know about these difficulties. To overcome them is a part of the scientific process but needs some organisation; we need interfaces between different branches. It is relevant that one must not take the existence of these interfaces for granted, but their construction can be part of a project. The critical aspect is the relation between model variables and measurable observables. The potential tension between model variables and observables becomes even more critical if one works not only in an inter- but in a transdisciplinary context. Besides the fact that one can no longer rely on well-established conventions of scientific argumentation and practices, questions of language and communication become demanding tasks that may require using narratives as tools for conveying content. In Sec. 5, we discussed opportunities and potential problems one faces when working with narratives. It is crucial to realise an essential distinction between art and the natural sciences if one does so. Art inspires and conveys a mood, whereas the natural sciences explain and transport facts. The boundary between art and natural sciences is, at a closer inspection, quite blurry. Nevertheless, if one uses storytelling methods as part of complex decision support processes, one must keep in mind the different objectives of art and science.

Finally, request R3 asks to clarify the responsibility of the reader. The systemic nature of the use of models requires considering the different types of users, as discussed in Sec. 2–4. In Sec 5, we embedded these arguments in a discussion that emphasises the role of communication. Notably, the "responsibility of the reader" is a concept that, in the context of computational modelling, does not only apply to individuals but should also be extended to institutions, especially the media. Responsible journalists (there are still many!) must criticise the results of computational models and develop an understanding of what a model can do.

As stated in the introduction, computational modelling probably faces historical opportunities. There is some loud mistrust of science, but it is a minority of the population that expresses it. The possibilities of data acquisition, computer technology, and a growing fundamental understanding of modelling offer the modellers the chance to have a tangible impact on society in fields ranging from personalised medicine over epidemiology and economic planning to climate change. However, the chance for impact brings the duty for responsible and ethical action. Acting responsible starts undoubtedly at the level of the individual scientist. In addition, we must implement social and administrative structures that allow the ethical use of computational models and actively promote them. In our view, it is crucial to recognise that such promotion must observe the lessons of transdisciplinarity and activate resources ranging from pure natural science over philosophy and cultural studies to art and politics.

Acknowledgements. This report brings together expertise from different research directions and projects. The coordination of efforts and the compilation of this report has been made possible by the project ACDC (Artificial Cells with Distributed Cores to Decipher Protein Function) that has been funded by the European Union's Horizon 2020 research and innovation programme under Grant Agreement No. 824060. Part of the presented context has been produced as contribution to the ACDC Summer School organised by the University of Trento and the Museo delle Scienze. The contributors to this report gathered at the WIVACE 2021 meeting in Winterthur, partially sponsored by the School of Engineering, Zurich University of Applied Sciences. Contributing expertise came from projects funded by the FOPH (Federal Office of Public Health) in Switzerland and Armasuisse.

References

1. Fleischmann, K., Wallace, W.: Ethical implications of computational modeling. The Bridge **41**(1), 45–51 (2017)
2. Wallace, W.A.: Ethics in Modeling. Emerald Group Publishing Ltd. (1994)
3. Shults, F.L., Wildman, W.J., Dignum, V.: The ethics of computer modeling and simulation. In: 2018 Winter simulation conference (WSC). IEEE (2018)
4. Noorman, M.: Computing and Moral Responsibility, in Stanford Encyclopedia of Philosophy (Spring 2020 Edition), E.N. Zalta, Editor. (2020)
5. Fellermann, H., et al.: The 2019 Conference on Artificial Life (full proceedings pdf). In: Artificial Life Conference Proceedings. MIT Press (2019)
6. Pfeifer, R., Scheier, C.: Understanding Intelligence. MIT press (2001)
7. Anderson, C.: The end of theory: the data deluge makes the scientific method obsolete. Wired Magazine **16**(7), 16-07 (2008)
8. Domingos, P.: The Master Algorithm: How the Quest for the Ultimate Learning Machine Will Remake Our World, Penguin. 322 (2015)
9. Gunning, D., et al.: XAI—Explainable artificial intelligence. Science Robotics **4**(37), eaay7120 (2019)
10. Ghassemi, M., Oakden-Rayner, L., Beam, A.L.: The false hope of current approaches to explainable artificial intelligence in health care. The Lancet Digital Health **3**(11), e745–e750 (2021)
11. Stauffer, D.: A biased review of sociophysics. J. Stat. Phys. **151**(1), 9–20 (2013)
12. Wigner, E.P.: The unreasonable effectiveness of mathematics in the natural sciences. Richard Courant lecture in mathematical sciences delivered at New York University, May 11, 1959. Communications on Pure and Applied Mathematics **13**: pp. 1–14 (1960)
13. Wolf, W.: The Ising model and real magnetic materials. Braz. J. Phys. **30**(4), 794–810 (2000)
14. Gnanvi, J., et al.: On the reliability of predictions on Covid-19 dynamics: A systematic and critical review of modelling techniques. Infectious Disease Modelling (2021)
15. Frigg, R., Hartman, S.: Models in Science, in Stanford Encyclopedia of Philosophy (Spring 2020 Edition), E.N. Zalta, Editor. (2020)
16. Tress, G., Tress, B., Fry, G.: Clarifying integrative research concepts in landscape ecology. Landscape Ecol. **20**(4), 479–493 (2005)
17. Klein, J.T.: A taxonomy of interdisciplinarity. The Oxford Handbook of Interdisciplinarity **15**, 15–30 (2010)
18. Fuchs, H.U.: From stories to scientific models and back: narrative framing in modern macroscopic physics. Int. J. Sci. Educ. **37**(5–6), 934–957 (2015)
19. Fuchs, H.U., et al.: How metaphor and narrative interact in stories of forces of nature. Narrative and Metaphor in Education. Look both ways. Routledge, London (2018)
20. Kenny, G.: What Are Your KPIs Really Measuring? Harvard Business Manager (2020)
21. Bak, M.A.: Computing fairness: ethics of modeling and simulation in public health. Simulation p. 0037549720932656 (2020)
22. Krütli, P., et al.: Prioritätenliste und Kontingentberechnung: Pandemievorbereitung in der Schweiz. (2018)
23. Barthes, R.: La mort de l'auteur. Manteia **5**, 12–17 (1968)

Correction to: Artificial Life and Evolutionary Computation

Johannes Josef Schneider, Mathias Sebastian Weyland,
Dandolo Flumini, and Rudolf Marcel Füchslin

Correction to:
J. J. Schneider et al. (Eds.): *Artificial Life and Evolutionary*
Computation, CCIS 1722,
https://doi.org/10.1007/978-3-031-23929-8

The following chapters were originally published electronically on the publisher's internet portal without open access:

"Obstacles on the Pathway Towards Chemical Programmability Using Agglomerations of Droplets", written by Johannes Josef Schneider, Alessia Faggian, Hans-Georg Matuttis, David Anthony Barrow, Jin Li, Silvia Holler, Federica Casiraghi, Lorena Cebolla Sanahuja, Martin Michael Hanczyc, Patrik Eschle, Mathias Sebastian Weyland, Dandolo Flumini, Peter Eggenberger Hotz, Rudolf Marcel Füchslin.

"The Good, the Bad and the Ugly: Droplet Recognition by a "Shootout"-Heuristics", written by Hans-Georg Matuttis, Silvia Holler, Federica Casiraghi, Johannes Josef Schneider, Alessia Faggian, Rudolf Marcel Füchslin, Martin Michael Hanczyc.

"Exploring the Three-Dimensional Arrangement of Droplets", written by Johannes Josef Schneider, Mathias Sebastian Weyland, Dandolo Flumini, Rudolf Marcel Füchslin.

"Geometric Restrictions to the Agglomeration of Spherical Particles", written by Johannes Josef Schneider, David Anthony Barrow, Jin Li, Mathias Sebastian Weyland, Dandolo Flumini, Peter Eggenberger Hotz, Rudolf Marcel Füchslin.

"Ethical Aspects of Computational Modelling in Science, Decision Support and Communication", written by Rudolf Marcel Füchslin, Jacques Ambühl, Alessia

---

The updated original version of these chapters can be found at
https://doi.org/10.1007/978-3-031-23929-8_4
https://doi.org/10.1007/978-3-031-23929-8_5
https://doi.org/10.1007/978-3-031-23929-8_6
https://doi.org/10.1007/978-3-031-23929-8_7
https://doi.org/10.1007/978-3-031-23929-8_24

J. J. Schneider et al. (Eds.): WIVACE 2021, CCIS 1722, pp. C1–C2, 2023.
https://doi.org/10.1007/978-3-031-23929-8_25

Faggian, Harold M. Fellermann, Dandolo Flumini, Armando Geller, Martin M. Hanczyc, Andreas Klinkert, Pius Krütli, Hans-Georg Matuttis, Thomas Ott, Stephan Scheidegger, Gary Bruno Schmid, Timo Smieszek, Johannes J. Schneider, Albert Steiner, Mathias S. Weyland.

With the authors' decision to opt for Open Choice the copyright of the chapters changed on 19 September 2023 to © Authors, 2023 and the chapters are forthwith distributed under a Creative Commons Attribution.

Funded by: the European Union's Horizon 2020 program "Artificial Cells with Distributed Cores to Decipher Protein Function" (ACDC), Grant Number: 824060.

Author Index

Printed in the United States
by Baker & Taylor Publisher Services